Lecture Notes in Computer Science 9792

Commenced Publication in 1973
Founding and Former Series Editors:
Gerhard Goos, Juris Hartmanis, and Jan van Leeuwen

Editorial Board

More information about this series at http://www.springer.com/series/7409

Alejandra González-Beltrán · Francesco Osborne
Silvio Peroni (Eds.)

Semantics, Analytics, Visualization

Enhancing Scholarly Data

Second International Workshop, SAVE-SD 2016
Montreal, QC, Canada, April 11, 2016
Revised Selected Papers

 Springer

Editors
Alejandra González-Beltrán
Oxford e-Research Centre
University of Oxford
Oxford
UK

Francesco Osborne
Knowledge Media Institute
The Open University
Milton Keynes
UK

Silvio Peroni
Department of Computer Science
and Engineering
University of Bologna
Bologna
Italy

ISSN 0302-9743   ISSN 1611-3349   (electronic)
Lecture Notes in Computer Science
ISBN 978-3-319-53636-1   ISBN 978-3-319-53637-8   (eBook)
DOI 10.1007/978-3-319-53637-8

Library of Congress Control Number: 2017933946

LNCS Sublibrary: SL3 – Information Systems and Applications, incl. Internet/Web, and HCI

Printed on acid-free paper

This Springer imprint is published by Springer Nature
The registered company is Springer International Publishing AG
The registered company address is: Gewerbestrasse 11, 6330 Cham, Switzerland

Preface

Supporting new forms of scholarly data publication and analysis is a crucial task for researchers, publishers, and companies working in innovative solutions for scholarly communication. Currently, most research papers are published as portable document format (PDF) and/or poorly annotated, only with simple metadata provided as a set of keywords or topic categories, if at all, which makes it hard to extract information from the full text. In addition, not always are other research outcomes, such as research data or software, made available nor do they include rich metadata. This hinders the discoverability, reuse, and reproducibility of research data and findings. A more structured and semantically rich representation of the research outcomes could bring significant advantages to various areas: linking more effectively research and industrial efforts, supporting researchers' work, fostering cross-pollination of ideas and methods across different areas, driving research policies, and acting as a source of information for a variety of applications. The program of the Semantics, Analytics, Visualization: Enhancing Scholarly Data (SAVE-SD) 2016 Workshop highlighted topics in these areas.

The first edition of the SAVE-SD workshop took place on May 19, 2015, and was co-located with the 24th International World Wide Web Conference in Florence, Italy. After the success of the first edition, the second edition, presented in this volume, took place on April 11, 2016, in Montreal, Canada, co-located with the 25th International World Wide Web Conference.

SAVE-SD 2016 opened with the keynote speech by Alex Wade, Director of Scholarly Communications at Microsoft Research, whose work focuses on Microsoft Academic and involves aspects of knowledge acquisition, knowledge representation, intentionality, dialog systems, semantic search, and intelligent agents. His talk was entitled "The Microsoft Academic Graph: New Applications and Research Opportunities" and described the new entity graph of research publications, authors, venues, organizations, and topics that is developed by Microsoft Research and drives new features in Bing, Cortana, and Microsoft Academic. We are grateful to Alex Wade for his inspiring talk.

The scientific program of SAVE-SD 2016 comprised 11 papers: five full papers, selected out of 11 submissions, which corresponds to an acceptance rate of 45%; two position papers and six poster or demo papers, selected out of five submissions, plus three of the full papers that instead were accepted as one position, one demo, and one poster paper. The workshop received a total of 16 submissions from authors of 15 countries in three continents (Europe, Asia, Americas) and was attended by about 50 people.

The topics in this edition demonstrate current research on semantic publishing and cover the extraction of semantic information from research papers or pre-existing datasets and the use of semantic techniques for characterizing citations and analyzing research topics and trends. SAVE-SD provided a forum for researchers, publishers, and companies interested in enhancing scholarly data to come together and discuss challenges and innovative solutions.

SAVE-SD accepts submissions not just in the traditional PDF format, but also in HTML, encouraging authors to provide semantically rich papers themselves. In order to support authors willing to submit in HTML, SAVE-SD encouraged the use of the RASH format (https://github.com/essepuntato/rash) for submissions. RASH stands for Research Articles in Simplified HTML (RASH) and can be produced from Open Office documents, Microsoft Word documents, and other formats.

SAVE-SD 2016 awarded two prizes: one for best paper and another for best RASH paper. The latter was sponsored by Springer Nature.

The criteria for selecting the best paper award considered the reviewers' scores and selected the paper with the best score. The best paper award was given to: *"Detection of Embryonic Research Topics by Analysing Semantic Topic Networks"* by Angelo Salatino and Enrico Motta.

The best RASH paper award is given to the paper that makes best use of the RASH format. This is chosen by an automatic score system that rates all the RASH submissions considering:

1. The quality of the markup (i.e., considering the number of errors in the document compared with the RASH grammar)
2. The quality of HTML (i.e., how many errors the document has compared with HTML5)
3. The number of Resource Description Framework (RDF) statements defined
4. The number of RDF links to Linked Open Data (LOD) datasets

The best RASH paper of SAVE-SD 2016 was awarded to: *"Citation Functions for Knowledge Export—A Question of Relevance, or, Can CiTO Do the Trick?"* by Joakim Philipson.

As SAVE-SD aims to address the gap between the theoretical and practical aspects of scholarly data, the review process ought to consider both perspectives. Thus, SAVE-SD has three different Program Committees (PCs):

– An Industrial PC, which mainly evaluates the submissions from an industrial perspective by assessing how much the theories or applications described in the papers (may) influence (positively or negatively) the publishing and technological domain and whether they could be concretely adopted by publishers and scholarly data providers
– An Academic PC, which evaluates the papers mainly from an academic perspective by assessing the quality of the research described in such papers
– A Senior PC, whose members act as meta-reviewers and have the crucial role of balancing the scores provided by the reviews from the other two PCs

We are very grateful to all the members of the three PCs, listed herein, for their high-quality reviews and constructive feedback, which improved significantly the quality of the papers contained in these proceedings.

Last but certainly not least, we want to thank our sponsors:

– Springer Nature (http://www.springernature.com/), who provided a 150-euro voucher to buy Springer Nature's products for the best RASH paper award

- Pensoft Publishers (http://www.pensoft.net/), who hosted a free-of-charge special collection for selected position/poster/demo papers in the *Research Idea and Outcomes* (RIO) journal
- *GigaScience* journal (http://www.gigasciencejournal.com/), which provided cool "Data is coming" t-shirts for the workshops attendees

March 2017

Alejandra González-Beltrán
Francesco Osborne
Silvio Peroni

Organization

Workshop Chairs

Alejandra González-Beltrán   Oxford e-Research Centre, University of Oxford, UK
Francesco Osborne   Knowledge Media Institute, Open University, UK
Silvio Peroni   Department of Computer Science and Engineering,
     University of Bologna, Italy

Senior Program Committee

Timothy W. Clark   Harvard University, USA
Aldo Gangemi   Université de Paris 13, France, and CNR, Italy
Ivan Herman   Digital Publishing Lead, W3C
Pascal Hitzler   Wright State University, USA
Enrico Motta   KMi, The Open University, UK
Susanna-Assunta Sansone   University of Oxford and NPG, UK
Daniel Schwabe   Pontifical Catholic University of Rio de Janeiro, Brazil
Simone Teufel   University of Cambridge, UK
Fabio Vitali   University of Bologna, Italy

Industrial Program Committee

Aliaksandr Birukou   Springer Nature
Scott Edmunds   GigaScience and BioMed Central
Anita de Waard   Elsevier
Patricia Feeney   CrossRef
Maarten Fröhlich   IOS Press
Paul Groth   Elsevier Labs
Laurel L. Haak   ORCID
Thomas Ingraham   F1000Research
Kris Jack   Mendeley
Petr Knoth   Mendeley
Michele Pasin   Springer Nature
Lyubomir Penev   Pensoft Publishers
Eric Prud'hommeaux   W3C
Anna Tordai   Elsevier
Alex Wade   Microsoft Research

Academic Program Committee

Andrea Bonaccorsi   University of Pisa, Italy
Davide Buscaldi   Université de Paris 13, France

Paolo Ciancarini	University of Bologna, Italy
Paolo Ciccarese	Harvard University, USA
Oscar Corcho	UPM, Spain
Mathieu d'Aquin	KMi, The Open University, UK
Rob Davey	Genome Analysis Centre, UK
Stefan Dietze	L3S Research Center, Germany
Angelo Di Iorio	University of Bologna, Italy
Alexander García Castro	Florida State University, USA
Leyla Jael García Castro	University of Munich, Germany
Daniel Garijo	UPM, Spain
Anna Lisa Gentile	University of Mannheim, Germany
Andrea Giovanni Nuzzolese	STC-CNR Rome, Italy
Asunción Gómez Pérez	UPM, Spain
Tudor Groza	Garvan Institute of Medical Research, Australia
Tom Heath	Open Data Institute, UK
Rinke Hoekstra	VU Amsterdam, The Netherlands
Tomi Kauppinen	Aalto University, Finland, and University of Münster, Germany
Steffen Lohmann	University of Stuttgart, Germany
Eamonn Maguire	CERN, Switzerland
Steve Pettifer	University of Manchester, UK
Francesco Poggi	University of Bologna, Italy
Philippe Rocca-Serra	University of Oxford, UK
Francesco Ronzano	Universitat Pompeu Fabra, Spain
Bahar Sateli	Concordia University, Canada
Jodi Schneider	University of Pittsburgh, USA
Ilaria Tiddi	KMi, The Open University, UK

Contents

Introduction

Alejandra González-Beltrán[1]([⊠]), Francesco Osborne[2], and Silvio Peroni[3]

[1] Oxford e-Research Centre, University of Oxford, Oxford, UK
alejandra.gonzalezbeltran@oerc.ox.ac.uk
[2] Data Science Group, Knowledge Media Institute,
The Open University, Milton Keynes, UK
francesco.osborne@open.ac.uk
[3] Digital and Semantic Publishing Laboratory,
Department of Computer Science and Engineering,
University of Bologna, Bologna, Italy
silvio.peroni@unibo.it

Abstract. We provide the motivation and overview of the SAVE-SD workshop series and introduce the manuscripts that were accepted as part of these proceedings of the SAVE-SD 2016 edition.

Keywords: SAVE-SD 2016 · WWW 2016 · Scholarly data · Semantics · Analytics · Visualisation

1 SAVE-SD Workshops

Research on supporting new forms of scholarly data publication and analysis has attracted high priority attention from both the industrial and the academic worlds. Funding agencies and publishers are now supporting open and accessible data publication – some evidence of that it is one of the major themes in the EU Horizon2020 program[1]. International community forums are also working in this direction. FORCE11[2] — a community of scholars, librarians, archivists, publishers and research funders — have the aim of facilitating knowledge creation and sharing. Similarly, the Research Data Alliance (RDA)[3], a community including to 4300 members from 111 countries (as of September 2016)[4], is working since 2013 on building the social and technical infrastructure to enable open data sharing.

At the same time, companies are becoming increasingly active in providing novel and more efficient ways to share and analyse research knowledge. Journals such as Springer Nature Scientific Data[5] and Oxford University Press Giga-

---

[1] Open Research Data in H2020: http://ec.europa.eu/research/press/2016/pdf/opendata-infographic_072016.pdf.
[2] The Future of Research Communications and e-Scholarship: http://www.force11.org/.
[3] Research Data Alliance: https://www.rd-alliance.org/.
[4] https://rd-alliance.org/node/51727.
[5] Scientific Data journal: http://nature.com/sdata/.

© Springer International Publishing AG 2016
A. González-Beltrán et al. (Eds.): SAVE-SD 2016, LNCS 9792, pp. 1–8, 2016.
DOI: 10.1007/978-3-319-53637-8_1

Science[6], among others, offer incentives for data publication and sharing. Thomson Reuters[7] and Elsevier[8] offer access to large datasets of scholarly data as a service to university and companies. Google Scholar[9] allows users to browse the large repository of paper indexed by Google. Microsoft Academic Search[10] offers a system for browsing research data and the Microsoft Academic Search Graph, a structured dataset containing metadata on research publications, authors, venues, organisations, and topics. Repositories supporting data publication and preservation (e.g., Zenodo[11], Dryad[12] and Figshare[13]) allow to make publicly available documents, datasets and other files in a citable, shareable and discoverable manner. Research social networks (ResearchGate[14], Academia.edu[15]) allow researchers to share and discuss their work with colleagues from all over the word. Altmetrics[16] and ImpactStory[17] offer services based on the analysis of social network for computing alternative metrics with the aim of assessing academic performance. Finally, a number of companies in the field of innovation brokering and "horizon scanning" (e.g. Idex Labs[18], Linknovate[19]) constantly analyse the research landscape for finding relevant experts and informing the strategies of client companies. Hence, the interest from the business world presents an unprecedented opportunity for rapidly transforming academic knowledge into practice and achieving the data-driven science and innovation that is promised by the Big Data era.

With respect to academia, several conferences related to the World Wide Web and the Semantic Web offered a number of workshops on related topics – such as:

1. SePublica 2011–2016 on semantic publishing at European Semantic Web Conference[20],
2. BigScholar 2014–2016 at the International World Wide Web Conference (on exploration and management of the Web of Scholars),
3. Linked Science 2011–2016 at the International Semantic Web Conference (on the use of Semantic Web technologies for integrating scientific data) and
4. the previous edition of this workshop, SAVE-SD 2015 at the International World Wide Web Conference.

---

[6] GigaScience journal: http://www.gigasciencejournal.com/.
[7] http://wokinfo.com/.
[8] http://www.scopus.com/.
[9] http://scholar.google.com/.
[10] http://academic.research.microsoft.com/.
[11] https://zenodo.org/.
[12] https://datadryad.org/.
[13] http://figshare.com/.
[14] http://researchgate.net/.
[15] https://www.academia.edu/.
[16] http://altmetrics.org/.
[17] https://impactstory.org/.
[18] https://www.idexx.com/.
[19] http://www.linknovate.com/.
[20] European Semantic Web Conference: http://www.eswc-conferences.org/.

We have also observed active participation in challenges such as the ESWC Semantic Publishing Challenge 2015[21] and 2016[22].

However, despite the rapid developments in this area, there is still a need for further dialogue between academia and industry, as well as other stakeholders working towards the vision of enhanced research data. In particular, the exchange of knowledge between the communities of *scholarly data representation, research data analytics, human computer interaction* and *visualisation* is still lacking. Indeed, research data need to be first annotated and enhanced semantically, then analysed, indexed, classified, enriched and visualised, and finally the resulting structured information should be conveyed to different kind of stakeholders in a user friendly and intuitive manner. Starting a dialogue between the experts in areas such as Knowledge Engineering, Semantic Web, Natural Language Processing (NLP), Scholarly Communication, Bibliometrics, Human-Computer Interaction, Information Visualisation is thus vital for realising a comprehensive workflow for sharing scientific knowledge.

Thus, the aim of the SAVE-SD workshops is to offer a forum to bring together researchers, publishers and other companies, to discuss the present scenarios concerning the production and use of scholarly data, and to strategise future research and industrial directions. We believe that the combination of different expertise and perspective could be a fertile ground for the creation of innovative and scalable solutions for sharing, reusing and processing research knowledge.

In particular, The SAVE-SD Workshops focus on the following topics:

1. *semantics* of scholarly data, i.e. how to categorise, connect, integrate and represent scholarly data and its provenance information semantically, in order to foster data sharing, interoperability, reusability and reproducibility;
2. *analytics* on scholarly data, i.e. designing and implementing novel and scalable algorithms for knowledge extraction with the aim of understanding research dynamics, forecasting research trends, fostering connections between groups of researchers, informing research policies, analysing and interlinking experiments and deriving new knowledge;
3. *visualisation* of and interaction with scholarly data, i.e. providing novel user interfaces and applications for navigating and making sense of scholarly data and highlighting their patterns and peculiarities.

This article introduces SAVE-SD 2016 proceedings, which corresponds to the second edition of the workshop. A selection of the papers from the first edition was published in the PeerJ Computer Science Journal[23].

2 SAVE-SD Advocating Enhanced Papers

The adoption of Web-based formats in scientific literature is an important step towards the complex and exciting vision of Semantic Publishing. The goal is

---

[21] https://github.com/ceurws/lod/wiki/SemPub2015.
[22] https://github.com/ceurws/lod/wiki/SemPub2016.
[23] https://peerj.com/collections/24-save-sd-2015/.

to unlock the knowledge hidden in other formats. For this reason SAVE-SD is actively encouraging author to submit their research papers in HTML-based formats.

In particular, SAVE-SD offers support for submission in the Research Articles in Simplified HTML (RASH) format[24]. RASH is a markup language that restricts the use of HTML to 32 elements [3]. This solution allows authors to include semantic relationships in their work either by associating RDFa (Resource Description Framework in Attributes) annotations or by inserting plain Turtle, RDF/XML (Resource Description Framework eXtensible Markup Language serialisation) or JSON-LD (Javascript Object Notation for Linking Data) content in a `script` element. To encourage submission in RASH the organisers introduced a special award for the best submission in RASH.

SAVE-SD 2015 was the first workshop to accept RASH papers. Currently, together with other HTML-based formats, RASH is accepted by the main Semantic Web conferences (ISWC[25], ESWC[26], EKAW[27]) and by a number of related workshops and challenges[28].

3   Short Overview of the Papers

While the papers contain to a degree all the components suggested by the SAVE-SD workshop, i.e., semantics, analytics or visualisation, we can classify them in two main categories: the manuscripts that address the extraction of semantic information from full-text or pre-existent datasets and the ones that focus on exploiting semantic techniques for fostering the analysis of citations, researchers and topics.

The *first category* includes three full papers, one position paper and five poster/demo papers.

In recent years there has been a number of efforts to extract scientific artefacts (e.g., genes [1], chemical components [2]) and epistemological concepts (e.g., hypothesis, motivation, experiments) [4,5] from research publications. The following five papers are dedicated to this fascinating task.

The paper by *Ronzano and Saggion* introduces a platform to represent in RDF several aspects of scientific publications, using techniques such as rhetorical sentence classification and text summarisation. The research publications are analysed by relying on the Text Mining Framework developed in the context of the European Project Dr. Inventor [8]. In line with the SAVE-SD vision, this framework also offers a number of relevant web visualisations[29] for exploring the produced RDF dataset.

---

[24] https://github.com/essepuntato/rash.
[25] International Semantic Web Conference: http://swsa.semanticweb.org/content/international-semantic-web-conference-iswc.
[26] European Semantic Web Conference: http://www.eswc-conferences.org/.
[27] Extended Knowledge Acquisition Workshop: http://ekaw.org/.
[28] https://github.com/essepuntato/rash/#venues-that-have-adopted-rash-as-submission-form
[29] http://backingdata.org/dri/viz/.

Gábor, Zargayouna, Tellier, Buscaldi and Charnois propose a method for automatically extracting semantic relations from articles in the science/engineering domain in their poster paper. Their approach allows to identify the entities and concepts that describe a scientific field (e.g., methods, problems) and the semantic relations between them (e.g., tackle, develop). The proposed workflow combines natural language processing techniques with statistical term extractors and external ontological resources.

Marsi and Øzturk's poster introduces a framework for finding events in natural science literature, such as the increase/decrease of variables. The resulting knowledge base enables semantic search for events and variables, which can be used to assess possible correlations – e.g., the increase in the sea level vs the decrease of the ice sheet. The system offers also a user interface to browse the events and visualise their type, frequency and relation strength.

Similarly, *Sateli and Witte*'s demo paper describes a workflow for converting a research paper in a Linked Open Data (LOD) compliant knowledge base. Their solution includes a Natural Language Processing (NLP) pipeline for tokenisation, sentence splitting, part-of-speech (POS) tagging, stemming, and verb group analysis; the Rhetector component for automatically detecting rhetorical entities; LODtagger for linking Dbpedia entities to the paper; and LODeXporter for generating the output RDF.

Finally, the poster paper by *Alexiou, Vahdati, Lange, Papastefanatos and Lohmann* presents the OpenAIRE LOD services, the RDF version of the well known Open Access Infrastructure for Research in Europe dataset[30], which includes publications and datasets from more than 100,000 research projects. In particular, the poster describes the scalable workflow used for the RDFization process of such a huge database.

The complex research entities extracted by these approaches have the potential to revolutionise the way we analyse scientific literature. However, key phrases are still the most common means to represent the content of articles for the benefits of users, search engine and recommendation systems. For this reason, the paper by *Daudaravicius* introduces a new statistical approach for extracting key phrases from scientific journals in the fields of astrophysics, mathematics, physics, and computer science. Their method uses the additive smoothing of term frequency-inverse document frequency (TF-IDF) for improving the quality of key phrases derived from large sample of papers.

References are one of the most used kinds of entities for understanding the research landscape, but extracting and visualising them is still a challenge. The demo paper by *Požega, Poljak and Kocijan* presents an approach which represents references from scientific papers as a tree-shape ReferenceTree model. The visualisation integrates multiple dimensions related to bibliographic references into the tree and allows one to easily browse an author's publication history.

The task of extracting semantic information from scholarly papers was addressed since 2014 by the Semantic Publishing Challenge (SemPub) at the Extended Semantic Web Conference. SemPub created a framework for compar-

---

[30] https://www.openaire.eu/.

ing in an objective way a number of systems in the semantic publishing domain and encouraged researcher to produce and make available a number of relevant Linked Data dataset. The paper of *Vahdati, Dimou, Lange and Di Iorio* examines the overall organisation of the Challenge and the results produced in the 2015 and 2016 editions. It also analyses the different system proposed for the different tasks and discusses a number of good lessons learned by the organisers.

Scholarly metadata can also be found on the web, in formats such as RDFa[31], Microdata[32] and Microformats[33]. However, it is not always easy to recover and aggregate this data. The position paper by *Sahoo, Gadiraju, Yu, Saha and Dietze* contributes to this challenge by presenting an analysis on Web Data Commons (WDC) dataset[34] with the aim of identifying frequent types and terms, the key providers of bibliographic markup and the most common errors. The findings include the prevalence of statements describing authors, publishers and keywords and the fact that Springer.org appears to be the most active data provider by a large margin in the sample under analysis.

The *second category* of papers addresses the use of semantic technologies for citation and topic analysis and is composed by two full papers, a position paper and a poster/demo paper.

The position paper by *Philipson* addresses the use of citation functions for promoting knowledge export and discusses the use of the Citation Typing Ontology (CiTO) [7] for this task. Indeed, while in many contexts different kinds of citations are treated as equal, they can be radically different according to their semantics and rhetorical context. The paper examines in particular cross-disciplinary citation functions, such as "comparison", "evidence", "force", "method" and "result". It concludes that currently CiTO is not specific enough to capture the subtle differences between some of citation functions and suggest that a combination of citation functions and subject headings, extracted from both citing and cited entities might offer even better prospects for knowledge export.

The rest of the papers highlight the advantage of a semantic characterisation of research topics [6] for describing researchers and analysing the evolution of research trends.

Sateli, Löffer, König-Ries and Witte propose a novel method for automatically creating authors' profiles according to their topics of interest. Indeed, a number of scholarly applications build on a representation of researchers in term of their competence, for supporting services such as expert search and paper recommendation. The automatically extraction of this profile from the full-text of research papers is performed by means of a text mining pipeline, which detects relevant topics as grounded named entities from DBpedia[35]. Interestingly, the

---

[31] http://www.w3.org/TR/xhtml-rdfa-primer/.
[32] http://www.w3.org/TR/microdata.
[33] http://microformats.org/.
[34] http://webdatacommons.org/.
[35] http://dbpedia.org/.

evaluation showed that the topic extracted within specific rhetorical zones are more representative of the author's competences.

The paper by *Salatino and Motta*, which won the best paper award, focuses on the detection of embryonic research topics that can be used for anticipating future research trends. It theorises that the appearance of novel research areas is anticipated by specific dynamics between existing ones and suggests a method based on the analysis of 3-cliques for detecting these dynamics. The paper presents an experiment on a sample of 3 million research papers which confirms the hypothesis. The main finding is that the pace of collaboration in the subgraphs of topics that will give rise to a new research area is significantly higher than the one in the control group. This knowledge could foster a variety of methods for trend detection which currently focus on topics already associated with a label or a substantial number of documents.

Portenoy and West poster paper addresses a similar issue, proposing a new kind of visualisation for representing the evolution of a topic and its influence on other fields, according to the citations graph. Their application exploits hierarchical clustering techniques to partition the citation graph into clusters representing fields and subfields. A demo of this visualisation is publicly available at http://scholar.eigenfactor.org/fields.

4   Journal Issue for Extended Papers

The authors of full papers were invited to submit an extended version of their work to a special issue that will be published as part of the PeerJ Computer Science. The authors of position, demo, and poster papers of the workshop were invited to submit an extended version of their works to a special issue that will be published as part of the Research Ideas and Outcomes (RIO) Journal.

The reader will be able to find further information of such extended papers at http://cs.unibo.it/save-sd/2016/.

References

1. Carpenter, B.: LingPipe for 99.99% recall of gene mentions. In: Proceedings of the Second BioCreative Challenge Evaluation Workshop, vol. 23, pp. 307–309 (2007)
2. Corbett, P., Copestake, A.: Cascaded classifiers for confidence-based chemical named entity recognition. BMC Bioinform. **9**(11), 1 (2008)
3. Di Iorio, A., Nuzzolese, A.G., Osborne, F., Peroni, S., Poggi, F., Smith, M., Vitali, F., Zhao, J.: The RASH Framework: enabling HTML+RDF submissions in scholarly venues. In: 14th International Semantic Web Conference (2015)
4. Groza, T.: Using typed dependencies to study and recognise conceptualisation zones in biomedical literature. PLoS ONE **8**(11), e79570 (2013)
5. Liakata, M., Teufel, S., Siddharthan, A., Batchelor, C.R.: Corpora for the conceptualisation and zoning of scientific papers. In: Proceedings of the 2010 International Conference on Language Resources and Evaluation (2010)
6. Osborne, F., Motta, E.: Klink-2: integrating multiple web sources to generate semantic topic networks. In: Arenas, M., et al. (eds.) ISWC 2015. LNCS, vol. 9366, pp. 408–424. Springer, Cham (2015). doi:10.1007/978-3-319-25007-6_24

7. Peroni, S., Shotton, D.: FaBiO and CiTO: ontologies for describing bibliographic resources and citations. Web Semantics: Sci. Serv. Agents World Wide Web **17**, 33–43 (2012)
8. Ronzano, F., Saggion, H.: Dr. Inventor Framework: Extracting Structured Information from Scientific Publications. In: Japkowicz, N., Matwin, S. (eds.) DS 2015. LNCS (LNAI), vol. 9356, pp. 209–220. Springer, Cham (2015). doi:10.1007/978-3-319-24282-8_18

Extracting Knowledge from Research Publications

Knowledge Extraction and Modeling from Scientific Publications

Francesco Ronzano(✉) and Horacio Saggion

Natural Language Processing Group (TALN),
Universitat Pompeu Fabra, Barcelona, Spain
{francesco.ronzano,horacio.saggion}@upf.edu

Abstract. During the last decade the amount of scientific articles available online has substantially grown in parallel with the adoption of the Open Access publishing model. Nowadays researchers, as well as any other interested actor, are often overwhelmed by the enormous and continuously growing amount of publications to consider in order to perform any complete and careful assessment of scientific literature. As a consequence, new methodologies and automated tools to ease the extraction, semantic representation and browsing of information from papers are necessary. We propose a platform to automatically extract, enrich and characterize several structural and semantic aspects of scientific publications, representing them as RDF datasets. We analyze papers by relying on the scientific Text Mining Framework developed in the context of the European Project Dr. Inventor. We evaluate how the Framework supports two core scientific text analysis tasks: rhetorical sentence classification and extractive text summarization. To ease the exploration of the distinct facets of scientific knowledge extracted by our platform, we present a set of tailored Web visualizations. We provide on-line access to both the RDF datasets and the Web visualizations generated by mining the papers of the 2015 ACL-IJCNLP Conference.

Keywords: Scientific knowledge extraction · Knowledge modeling · RDF · Software framework

1 Introduction: Dealing with Scientific Publications Overload

Currently, researchers have access to a huge and rapidly growing amount of scientific literature available on-line. Recent estimates reported that a new paper is published every 20 s [1]. PubMed[1], the reference publication index for life

This work is (partly) supported by the Spanish Ministry of Economy and Competitiveness under the Maria de Maeztu Units of Excellence Programme (MDM-2015-0502) and by the European Project Dr. Inventor (FP7-ICT-2013.8.1 - Grant no: 611383).

[1] http://www.ncbi.nlm.nih.gov/pubmed.

© Springer International Publishing AG 2016
A. González-Beltrán et al. (Eds.): SAVE-SD 2016, LNCS 9792, pp. 11–25, 2016.
DOI: 10.1007/978-3-319-53637-8_2

science and biomedical topics, currently includes about 24.6 million papers with a growth rate of about 1,370 new articles per day. Elsevier' Scopus[2] and Thomson Reuter's ISI Web of Knowledge[3], the two biggest privately held journal indexes, respectively contain more than 57 and 90 million papers.

At the same time, during the last few years the number of scientific papers that are freely accessible on-line considerably grew [2,3]. Currently, the Directory of Open Access Journals[4], one of the most authoritative indexes of high quality, Open Access, peer-reviewed publications, lists more than 10,800 journals and 2.1 million papers. The full text of 27% of the articles indexed by PubMed is available on-line for free. Sometimes between 2017 and 2021, more than half of the global papers are expected to be published as Open Access articles [4].

The exploration of recent advances concerning specific topics, methods and techniques, peer reviewing, the writing and evaluation of research proposals and in general any activity that requires a careful and comprehensive assessment of scientific literature has turned into an extremely complex, time-consuming task.

In this context, considering also the increasing amount of scientific information freely accessible on-line, the availability of text mining tools able to extract, aggregate and turn scientific unstructured textual contents into well organized and interconnected knowledge is fundamental. However, scientific publications are characterized by several structural, linguistic and semantic peculiarities: general purpose text mining tools and techniques often need to be substantially adapted and extended in order to correctly deal with their contents. Even if the adoption of Web-friendly, textual formats and XML dialects like JATS[5] [5], Elsevier Schemas[6] and RASH[7] is rapidly spreading, *the majority of scientific papers is still available as PDF documents*, thus requiring proper tools to consistently extract their contents [6,8,9]. Scientific publications include *common structural elements* (title, authors, abstract, sections, figures, tables, citations, bibliography) that often requires customized approaches to be properly characterized [10–13]. Similarly, scientific articles are also distinguished by their *peculiar discursive structure* (background, challenge, outcome, future works) [14,15]. Papers are interconnected by their *network of citations* that constitutes the basis of widespread count-based metrics (i.e. h.index). Citation semantics has started to be exploited in several contexts including opinion mining [16,17] and scientific text summarization [18,19]. Integrated scientific article mining systems have been proposed and released in order to perform complex paper analysis tasks like the joint annotation of several kinds of structural information [25] or the semantic characterization and querying of contents [22].

---

[2] http://www.scopus.com/.

[3] http://www.webofknowledge.com/.

[4] https://doaj.org/.

[5] http://jats.nlm.nih.gov/.

[6] http://www.elsevier.com/author-schemas/elsevier-xml-dtds-and-transport-schemas.

[7] https://rawgit.com/essepuntato/rash/master/documentation/index.html.

Recently, in parallel to the diffusion of new approaches to scientific text mining, several investigation and development efforts have also been focused on the modeling and interlinking of scholarly publishing contents by relying on Semantic Web standards and technologies [20–22]. This trend is usually referred to as semantic publishing [23]. In this context, the Semantic Publishing Challenges [24], organized as part of the Extended Semantic Web Conferences, represents an important discussion and evaluation venue.

In this paper, we present a platform that extracts semantically rich information from scientific articles and represents it both as RDF datasets and by means of properly tailored Web visualizations. To mine the contents of scientific publications, we rely on the Text Mining Framework developed in the context of the European Project Dr. Inventor. The Framework integrates several text mining modules that spot many structural and semantic facets of scientific publications. In comparison with existing tools, the Dr. Inventor Text Mining Framework provides a coherent system that enables the automated extraction of a greater and richer set of structural and semantic knowledge facets from scientific articles. Besides the identification and enrichment of papers' structural and citation-related data, by relying on the Framework it is possible to perform the automated rhetorical classification of sentences, the disambiguation and entity linking of papers' contents and the creation extractive summaries of an article. Moreover it enables the creation of subject-verb-object graph representations of an article that are being exploited in the context of the Dr. Inventor Project to identify creative analogies across papers [39]. The Framework is distributed as a self-contained Java library[8], thus providing a convenient tool both to bootstrap more complex scientific publication analysis experiments as well as to foster the creation of structured, semantically-rich knowledge from papers' contents.

In Sect. 2 we describe the main scientific text mining modules that compose the Framework. Section 3 provides an evaluation of the performances of the Framework with respect to the identification of the discourse rhetorical category of sentences and the selection of the most relevant sentences to summarize a paper. In Sect. 4 we outline our approach to the representation of the contents mined from a paper as an RDF dataset. Section 5 introduces a set of Web visualizations useful to provide an easy and interactive way to explore the information extracted from a scientific publication. In Sect. 6 we present our conclusions and sketch future venues of research.

2   Exploiting the Dr. Inventor Framework to Mine Scientific Publications

We rely on the Dr. Inventor Text Mining Framework [26] (DRI Framework) to extract and enrich the information necessary to generate both RDF datasets and Web visualizations from scientific publications. The DRI Framework integrates, extends and customizes a collection of scientific text mining modules and services in order to support the joint analysis of structural, linguistic and semantic

---

[8] http://backingdata.org/dri/library/.

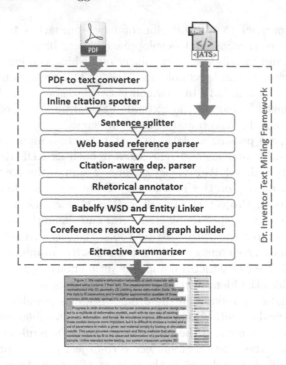

Fig. 1. Architectural overview of the modules of the Dr. Inventor Framework.

aspects of scientific publications. It has been implemented and is distributed as a Java library. The DRI Framework relies on the GATE Text Engineering Platform [27] as a common glue to integrate its text mining modules. Figure 1 provides an overview of the modules integrated in the current version of the DRI Framework. Each one of them is described in greater detail in the remaining part of this Section.

The DRI Framework can mine scientific publications both in PDF and JATS XML format. As shown in Fig. 1, two additional text mining modules are needed to process the contents of PDF articles, with respect to publications available as JATS XML files. The first module is the **PDF to text converter** that extracts textual contents from PDF documents. After a comparative analysis and evaluation of several PDF-to-text conversion approaches both generic and customized to scientific publications, we decided to rely on PDFX and its Web API[9] [6] to convert PDF files to text. PDFX is a rule-based PDF mining engine that enables most of the times the extraction of clean and consistent semi-structured textual contents form the PDF file of a scientific article. We rely on the structured XML output of PDFX to identify the title, the abstract, the sections and the bibliographic entries of a paper.

---

[9] http://pdfx.cs.man.ac.uk/.

Once PDF papers are converted to text, the **Inline citation spotter** module is executed. By means of a set of JAPE rules [37] covering several widespread citation styles, inline citation spans and inline citation markers are identified inside the textual contents of a paper (Fig. 2-a). Then each inline citation marker is linked to the related bibliographic entry (bibEntry) by a set of heuristics tailored to the detected inline citation style (Fig. 2-b).

The **PDF to text converter** module and the **Inline citation spotter** module are not needed for publications available as JATS XML files since their XML markup already identifies the structural elements just contemplated (sections, citations and related bibEntries).

The **Sentence splitter module** identifies the sentence boundaries inside each article by relying on a rule-based sentence splitting approach [27] that has been customized so as to deal with some peculiarity of scientific publications (expressions like: i.e., et. al., Fig., Tab. that do not identify the end of a sentence).

The **Web based reference parser** analyzes the contents of each bibEntry in order to identify its structural components (like paper title, authors, publication year, etc.). It also retrieves references to those bibEntries from external publication indexes (Fig. 2-c) by querying and merging the results of the on-line Web APIs exposed by *Bibsonomy*[10], *CrossRef*[11] and *FreeCite*[12].

At this stage, every sentence of the paper is processed by means of the next two modules. First of all the **Citation-aware dependency parser** performs the tokenization, lemmatization and POS-tagging of each sentence and builds a dependency tree by relying on a modified version of the MATE tools[13] [7] that has been properly customized to correctly deal with inline citation spans. When an inline citation span has a syntactic role inside the sentence where it occurs, it is considered as a single word when building the dependency tree of the sentence (Fig. 2-d, first example). On the contrary, when the inline citation span has not syntactic role in the sentence, it is ignored (Fig. 2-d, second example). The upper part of 2-e shows the POS tags and the dependency tree of a sentence in which the subject is the inline citation span *(Hu, 2004)*.

Thanks to the sentence analysis performed by the Citation-aware dependency parser, the **Rhetorical annotator** processes the contents of each sentence to identify its scientific discourse rhetorical category (see [29] for details on the annotation schema) among: Approach, Challenge, Background, Outcomes and Future Work. This module relies on a Logistic Regression classifier trained on the manual annotations of Dr. Inventor Corpus[14] [29,30]. In Sect. 3 we provide more details on the Corpus and evaluate the performance of this module.

The next module of the DRI Framework is the **Babelfy WSD and Entity Linker**. It processes the contents of the paper by invoking the Babelfy Web

---

[10] http://www.bibsonomy.org/help/doc/api.html.
[11] http://search.crossref.org/help/api.
[12] http://freecite.library.brown.edu/welcome.
[13] https://code.google.com/p/mate-tools/.
[14] http://sempub.taln.upf.edu/dricorpus.

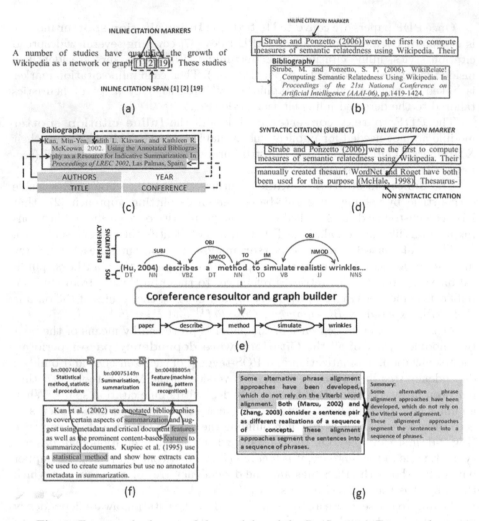

Fig. 2. Functional schemas of the modules of the Dr. Inventor Framework.

API[15] [33]. Babelfy is a graph-based methodology to perform Entity Linking and Word Sense Disambiguation, relying on the Babelnet semantic network[16]. Thanks to Babelfy the occurrences of concepts and Named Entities are spotted inside the text of each paper and properly linked to their right meaning chosen in the sense inventory of Babelnet. Figure 2-f shows a portion of an article where the occurrences of three concepts (*summarizaiton*, *features* and *statistical method*) have been spotted and linked to their respective Babelnet synsets (senses).

The **Coreference resolutor and graph builder** module, starting from the outputs of the Citation-aware dependency parser, represents each sentence

---

[15] http://babelfy.org/.
[16] http://babelnet.org/.

of a paper as a Subject-Verb-Object graph. An example of such graph is shown in Fig. 2-e. A rule-based nominal and pronominal coreference resolutor has been implemented in this module in order to support the integration of Subject-Verb-Object graphs generated from distinct sentences by merging the nodes that refer to the same entity. For instance, the coreference resolutor is able to spot that a pronominal node refers to a specific nominal entity, thus merging of both nodes.

The last module of the DRI Framework is the **Extractive summarizer**. It implements extractive paper summarization algorithms thanks to the integration of the SUMMA toolkit [34][17]. These algorithms rate the sentences of a paper with respect to their relevance for the inclusion in a summary: the top-n rated sentences are then chosen and composed so as to generate the extractive summary of the article (Fig. 2-g). The current version of the DRI Framework implements two basic sentence ranking approaches: the sentence similarity with the title of the paper and the sentence similarity with the centroid of each section of the paper. In Sect. 3 we evaluate the performance of distinct summarization approaches including the ones implemented by this module.

The DRI Framework is distributed as a self-contained Java library that exposes a convenient API in order to invoke the execution of the scientific text mining modules described in this Section. The results of the paper analyses can be easily accessed thanks to the tailored object-oriented data model of scientific publication that is implemented by the DRI Framework. The last version of the DRI Framework as well as the related JavaDoc, tutorials and code examples can be accessed online at: http://backingdata.org/dri/library/.

3   Evaluation of Rhetorical Sentence Annotation and Extractive Summarization

In this Section we present two experiments useful to measure the performance of two core modules of the DRI Framework: the Rhetorical annotator and the Extractive summarizer. Both experiments rely on the textual annotations of the Dr. Inventor Corpus. This Corpus includes 40 Computer Graphics papers containing 8,877 sentences that have been manually annotated with respect to their scientific discourse rhetorical category. Moreover, the corpus includes for each paper three handwritten summaries of maximum 250 words.

The **Rhetorical annotator** module integrated in the DRI Framework is based on a Logistic Regression rhetorical sentence classifier implemented by relying on the Weka data mining tools [38]. To select the best approach to determine the rhetorical category of each sentence, we compared the performance of two classifiers: Support Vector Machine (SVM) with linear kernel [28] and Logistic Regression. We represent each sentence to classify by means of a set of lexical and semantic features and evaluate each classification approach by performing a 10-fold cross validation over the 8,877 manually annotated sentences of Dr. Inventor Corpus [29]. The results are shown in Table 1 where we can notice that

---

[17] http://www.taln.upf.edu/pages/summa.upf/.

Table 1. F1 score of Logistic Regression and SVM classifier evaluated by a 10-fold cross validation over the manually annotated sentences of Dr. Inventor Corpus.

Rhetorical category	Logistic regression	SVM
Approach	0.876	0.851
Background	0.778	0.735
Challenge	0.466	0.430
Future work	0.675	0.496
Outcome	0.679	0.623
Avg. F1:	**0.801**	**0.764**

the Logistic Regression performs better than the SVM classifier both on average and with respect to each rhetorical category. In general, the performance of the classifier for each rhetorical category decreases with respect to the frequency of annotated sentences belonging to that category in the Dr. Inventor Corpus.

The **Extractive summarizer** module implements distinct approaches to rank the sentences of a paper with respect to their relevance to be included in a summary. In the rest of this Section we compare the summarization performances three approaches: the two summarization techniques implemented by the DRI Framework (sentence similarity with the title of the paper and sentence similarity with the centroid of each section of the paper) and the TextRank graph-based summarization algorithm [31].

To this purpose, for each paper of Dr. Inventor Corpus we generate three summaries of approximately 250 words, each one by relying on a specific summarization approach. Then, we compare each automatically generated summary with the three human handwritten ones by computing the average ROUGE-2 score[18] [32]. For each summarization approach, we determine the global ROUGE-2 score by computing the average ROUGE-2 of all the 40 papers of Dr. Inventor Corpus. In this way we can quantify and compare the performance of each summarization approach. By scoring sentences with respect to their similarity with the title, we obtain a global ROUGE-2 score of 0.3151 that improves up to 0.3427 when we score sentences by considering their similarity with each section centroid. The best summarization performance (global ROUGE-2 0.3617) is obtained by relying on the TextRank algorithm. We are planning to integrate this algorithm in the next releases of the DRI Framework.

4   Semantic Modeling of Scientific Publications

In this Section we present our approach to model as RDF data the structural and semantic information mined by means of the DRI Framework, thus enabling the automated creation of structured, rich data collections describing scholarly

---

[18] Rouge-2 is a measure which compares n-grams in automatic summaries to n-grams in gold stadard summaries.

contents, in accordance to the semantic publishing principles. We extend and enrich the basic RDF data modeling approach of scientific papers we adopted in the context of our participation to the Semantic Publishing Challenge 2015 [35]. In particular, our RDF data modeling choices have been driven by the necessity to represent the varied information that can be mined from a publication thanks to the DRI Framework. Thus, besides the representation of articles' metadata and bibliographic entries, our RDF data model contemplates the possibility to describe the structure of a paper, by identifying its abstract, sections and sentences. Each sentence can be characterized by both its rhetorical category and the Babelnet synsets (senses) that have been spotted inside its content. Moreover we link each bibEntry to all the sentences that include the related in-line citations.

The DRI Framework Java library has been properly extended with methods useful to trigger the automatic generation of the RDF dataset of a paper. The RDF datasets generated from the papers presented at the 2015 ACL-IJCNLP Conference can be downloaded online[19]. In the remaining part of this Section we describe in more detail the RDF data modeling choices we made and the ontologies we reused and extended to represent the contents of scientific publications.

Figure 3 schematizes our RDF model of scientific articles. We relied on the core RDF data modeling approaches, patterns and ontologies accessible in the Semantic Publishing and Referencing (SPAR) Portal[20] [36]. The SPAR Portal defines and documents a complete and consistent set of 12 ontologies tailored to model several aspect of scientific publishing, including articles' metadata, authors, bibliography, citations, publication workflows, etc. From the classes and the properties modeled by the SPAR ontologies, we reused and derived - in the dri namespace - new sub-classes and sub-properties. As a consequence, we included the related T-BOX axioms in the RDF Datasets we generate. The URIs needed to unambiguously reference each article together with its components (authors, sections, sentences, bibEntries, etc.) are instantiated in a namespace provided by DRI Framework users.

Figure 3-a shows how we represent the structured contents of a paper as RDF triples. Two URIs are generated to reference the abstract and the body of the paper (respectively the *FrontMatter_URI* and the *BodyMatter_URI* in Fig. 3-a). Both the abstract and the body may contain a list of sections (*IntroSection_URI* and *MethodSection_URI* in Fig. 3-a). Each section is identified by an URI and related to an instance of the doco:SectionTitle class that represents its title. The abstract, body or sections of the paper can contain one or more sentences, each one identified by an URI (*Sentence1_URI*, ..., *SentenceN_URI* in Fig. 3-a). The lower part of Fig. 3-a shows the association of the sentences of the paper to their scientific discourse rhetorical category. This is achieved by representing the corresponding *sentence_URI* as an instance of one of the following classes: dri:Approach, dri:Challenge, dri:Background, dri:Outcomes and dri:FutureWork. The association of a Babelnet synset (sense) to the sentence where the same

---

[19] Download link: http://backingdata.org/dri/viz/.
[20] http://www.sparontologies.net/.

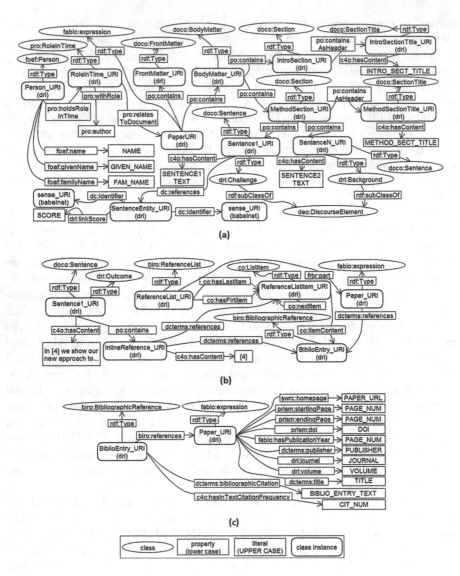

Fig. 3. RDF data model of scientific article: (a) authors and internal structure of the paper including sections and sentences with their rhetorical class and associated Babelnet senses; (b) list of bibliographic entries of the paper together with the pointer to the sentences in which each bibliographic entry occurs; (c) descriptive data of both papers and bibliographic entries. Ontology prefixes: **doco** *Document Components Ontology*, **fabio** *FRBR-aligned Bibliographic Ontology*, **c4o** *Citation Counting and Context Characterization Ontology*, **pro** *Publishing Roles Ontology*, **biro** *Bibliographic Reference Ontology*, **swrc** for the Semantic Web for Research Communities Ontology, **prism** *PRISM Metadata Ontology*, **foaf** *Friend Of A Friend Ontology*, **po** *Pattern Ontology*, **co** *Collections Onology*, **dc** and **dcterms** *Dublin Core Ontology*. The prefix **dri** identifies the classes and properties of Dr. Inventor Ontology.

synset has been spotted is modeled by linking the URI of the sentence to a *SentenceEntity_URI*. The *SentenceEntity_URI* is in turn characterized by the URIs of both the Babelnet synset and the DBpedia entity that represent that sense. Moreover, each association of a sense to a sentence is characterized by a score (literal object of the property dri:linkScore). This score is a double value that provides an estimate of the strength of the concept-to-sentence association.

On the left side of Fig. 3-a, we show how the Publishing Roles Ontology is exploited in order to model the authors of a paper. The same ontology is also used to represent the editors of an article.

Figure 3-b and c show the RDF representation of the bibliography of a paper. By relying on the Collections Ontology, the bibEntries are represented as an ordered list. An URI is assigned to each inline citation belonging to a specific sentence of the paper (*InlineRefrence_URI* in Fig. 3-b). The inline reference URI relates the sentence that contains the inline citation to the referenced bibEntry. Also the textual contents of the inline reference are specified by means of the property c4o:hasContent. Figure 3-c shows how each bibEntry is characterized by specifying the cited paper (identified by its URI, *Paper_URI*), the text of the same bibEntry and the number of times that bibEntry is cited inside the considered paper.

When we generate these data our focus is put on the creation of a consistent and semantically-rich RDF representation of the contents mined from a single scientific publication by means of the DRI Framework. As far as concern the creation of links to external Linked Data, the RDF datasets we generate connect publications and bibEntries to bibliographic indexes like Bibsonomy and relates each sentence of the paper to the Babelnet synsets (senses) mentioned in its contents. We plan to extend our RDF generation approach so as to foster the creation of new, richer internal and external links, thus increasing data integration and interlinking.

5   Visualizing Semantically Enriched Scientific Publications

In this Section we present a set of Web visualizations we developed to support an easier and more interactive navigation of the contents mined from a scientific publication by means of the DRI Framework. The visualizations of the papers presented at the 2015 ACL-IJCNLP Conference can be accessed online[21].

The information mined from a scientific article is presented by means of a multi-tab view (see Fig. 4). Each tab is meant to show a specific type of data extracted from a scientific publication together with aggregated statistical information. In the first tab, named 'Main tab' and shown in Fig. 4-a, the textual content of the paper can be browsed by section. Inline citations inside each sentence of the paper can be inspected (by a click) so as to explore the detailed metadata associated by the **Web based reference parser** module. Moreover,

---

[21] http://backingdata.org/dri/viz/.

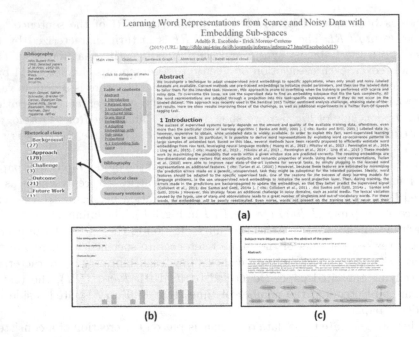

Fig. 4. Web visualizations of the information mined by the DRI Framework: (a) Main tab; (b) Citation tab; (c) Abstract graph tab.

the sentences of the paper can be highlighted in different colors with respect to the scientific discourse rhetorical category associated by the **Rhetorical annotator** module. Similarly it is also possible to highlight the sentences chosen by the **Extractive summarizer** to be part of a summary of the paper. All these features of the 'Main view' tab can be accessed by the four drop down menus that are present in its left side (Fig. 4-a).

The second tab, named 'Citations' (Fig. 4-b), enables the visualization of several aggregated statistical data concerning the citations of the paper. The third and the fourth tabs enable the visualization of the Subject-Verb-Object graphs (see Fig. 2-e) that represent respectively the contents of each sentence of the paper ('Sentence graph' tab) and the aggregated contents of the abstract of the paper ('Abstract graph' tab, Fig. 4-c). The Subject-Verb-Object graphs are mined by both the **Citation-aware dependency parser** module and the **Coreference resoultor and graph builder** module. A fifth tab named 'Babel senses cloud' enables users to inspect the top-10 Babelnet synsets (senses) that occur in the contents of the paper, identified thanks to the **Babelfy WSD and Entity Linker** module.

6   Conclusions and Future Work

The amount of scientific publications available on-line is growing at an unprecedented rate together with the diffusion of the Open Access publishing model, thus turning any careful and comprehensive assessment of scientific literature into an extremely complex and time-consuming task. In this scenario, in order to help researcher and other interested actors to easily select, access and aggregate the contents of scientific papers, the availability of new approaches and tools that enable the automated analysis and interconnection of structural and semantic information from scientific literature is fundamental.

In this paper we presented a platform useful to extract several types of information from scientific publications and represent it both as RDF datasets and by means of interactive Web visualizations. In order to process, analyze and enrich the contents of a scientific article we exploited the scientific Text Mining Framework we developed in the context of the European Project Dr. Inventor. We described in detail both the scientific text analysis modules integrated into the Framework and the RDF data modeling approach we adopted. We evaluated how the framework supports rhetorical sentence classification and extractive summarization. Moreover, we presented a set of Web visualizations of the structured contents we extract from scientific articles. The Dr. Inventor Text Mining Framework is available as a self-contained Java library that provides a comprehensive, ready-to-use platform for scientific text analysis. The Framework is intended to provide an integrated tool to ease the expensive and time consuming bootstrapping of scientific text mining experiments by automatically enriching the contents of scientific papers by identifying several structural and semantic information. The Framework is also meant to foster the automated creation of scholarly publishing RDF data since it allows the creation of RDF datasets that model the knowledge mined from a paper.

As future work, we plan to further improve and extrinsically evaluate the main text analysis modules of the Text Mining Framework. In particular we plan to refine and carry out user and task-based evaluations of the Subject-Verb-Object graphs extracted from the textual contents of each paper. We are also planning to experiment new approaches to rhetorical sentence classification by relying on active learning. We would like to evaluate new ways to further characterize and take advantage of the citations of a paper by determining their polarity and purpose.

References

1. Munroe, R.: The rise of open access. Science **342**(6154), 58–59 (2013). https://www.sciencemag.org/content/342/6154/58.full
2. Björk, B.C., Laakso, M., Welling, P., Paetau, P.: Anatomy of green open access. J. Assoc. Inf. Sci. Technol. **65**(2), 237–250 (2014)
3. Solomon, D.J., Laakso, M., Björk, B.C.: A longitudinal comparison of citation rates and growth among open access journals. J. Inf. **7**(3), 642–650 (2013)
4. Lewis, D.W.: The inevitability of open access. Coll. Res. Libr. **73**(5), 493–506 (2012)
5. Huh, S.: Coding practice of the journal article tag suite extensible markup language. Sci. Editing **1**(2), 105–112 (2014)

6. Constantin, A., Pettifer, S., Voronkov, A.: PDFX: fully-automated PDF-to-XML conversion of scientific literature. In: Proceedings of the 2013 ACM Symposium on Document Engineering, pp. 177–180. ACM (2013)
7. Bohnet, B.: Very high accuracy and fast dependency parsing is not a contradiction. In: Proceedings of the 23rd International Conference on Computational Linguistics, pp. 89–97. Association for Computational Linguistics (2010)
8. Tkaczyk, D., Szostek, P., Dendek, P.J., Fedoryszak, M., Bolikowski, L.: CERMINE-automatic extraction of metadata and references from scientific literature. In: 2014 11th IAPR International Workshop on Document Analysis Systems (DAS), pp. 217–221. IEEE (2014)
9. Ramakrishnan, C., Patnia, A., Hovy, E.H., Burns, G.A.: Layout-aware text extraction from full-text PDF of scientific articles. Sour. Code Biol. Med. **7**(1), 7 (2012)
10. Peng, F., McCallum, A.: Information extraction from research papers using conditional random fields. Inf. Process. Manage. **42**(4), 963–979 (2006)
11. Do, H.H.N., Chandrasekaran, M.K., Cho, P.S., Kan, M.Y.: Extracting and matching authors and affiliations in scholarly documents. In: Proceedings of the 13th ACM/IEEE-CS Joint Conference on Digital Libraries, pp. 219–228. ACM (2013)
12. Councill, I.G., Giles, C.L., Kan, M.Y.: ParsCit: an open-source CRF reference string parsing package. In: LREC (2008)
13. Luong, M.T., Nguyen, T.D., Kan, M.Y.: Logical structure recovery in scholarly articles with rich document features. In: Multimedia Storage and Retrieval Innovations for Digital Library Systems, vol. 270 (2012)
14. Liakata, M., Saha, S., Dobnik, S., Batchelor, C., Rebholz-Schuhmann, D.: Automatic recognition of conceptualization zones in scientific articles and two life science applications. Bioinformatics **28**(7), 991–1000 (2012)
15. Teufel, S.: The structure of scientific articles: applications to citation indexing and summarization. Comput. Linguist. **38**(2), 443–445 (2012)
16. Nakov, P.I., Schwartz, A.S., Hearst, M.: Citances: citation sentences for semantic analysis of bioscience text. In: Proceedings of the SIGIR 2004 Workshop on Search and Discovery in Bioinformatics, pp. 81–88 (2004)
17. Abu-Jbara, A., Ezra, J., Radev, D.R.: Purpose and polarity of citation: towards NLP-based bibliometrics. In: HLT-NAACL, pp. 596–606 (2013)
18. Abu-Jbara, A., Radev, D.: Coherent citation-based summarization of scientific papers. In: Proceedings of the 49th Annual Meeting of the Association for Computational Linguistics: Human Language Technologies, vol. 1, pp. 500–509. Association for Computational Linguistics (2011)
19. Ronzano, F., Saggion, H.: Taking advantage of citances: citation scope identification and citation-based summarization. In: Text Analytics Conference (2014)
20. Smit, E., Van Der Graaf, M.: Journal article mining: the scholarly publishers' perspective. Learn. Publ. **25**(1), 35–46 (2012)
21. Ciancarini, P., Iorio, A., Nuzzolese, A.G., Peroni, S., Vitali, F.: Semantic annotation of scholarly documents and citations. In: Baldoni, M., Baroglio, C., Boella, G., Micalizio, R. (eds.) AI*IA 2013. LNCS (LNAI), vol. 8249, pp. 336–347. Springer, Cham (2013). doi:10.1007/978-3-319-03524-6_29
22. Sateli, B., Witte, R.: What's in this paper?: Combining rhetorical entities with linked open data for semantic literature querying. In: Proceedings of the 24th International Conference on World Wide Web Companion, pp. 1023–1028 (2015)
23. Shotton, D.: Semantic publishing: the coming revolution in scientific journal publishing. Learn. Publ. **22**(2), 85–94 (2009)

24. Iorio, A.D., Lange, C., Dimou, A., Vahdati, S.: Semantic publishing challenge – assessing the quality of scientific output by information extraction and interlinking. In: Gandon, F., Cabrio, E., Stankovic, M., Zimmermann, A. (eds.) SemWebEval 2015. CCIS, vol. 548, pp. 65–80. Springer, Cham (2015). doi:10.1007/978-3-319-25518-7_6
25. Tkaczyk, D., Szostek, P., Fedoryszak, M., Dendek, P.J., Bolikowski, L.: CERMINE: automatic extraction of structured metadata from scientific literature. Int. J. Doc. Anal. Recogn. (IJDAR) 18(4), 317–335 (2015)
26. Ronzano, F., Saggion, H.: Dr. Inventor framework: extracting structured information from scientific publications. In: Japkowicz, N., Matwin, S. (eds.) DS 2015. LNCS (LNAI), vol. 9356, pp. 209–220. Springer, Cham (2015). doi:10.1007/978-3-319-24282-8_18
27. Cunningham, H., Tablan, V., Roberts, A., Bontcheva, K.: Getting more out of biomedical documents with GATE's full lifecycle open source text analytics. PLoS Comput. Biol. 9(2), e1002854 (2013)
28. Schölkopf, B., Smola, A.J.: Learning with Kernels: Support Vector Machines, Regularization, Optimization, and Beyond. MIT press, Cambridge (2002)
29. Fisas, B., Ronzano, F., Saggion, H.: On the discoursive structure of computer graphics research papers. In: The 9th Linguistic Annotation Workshop held in Conjuncion with NAACL 2015, p. 42 (2015)
30. Fisas, B., Ronzano, F., Saggion, H.: A multi-layered annotated corpus of scientific papers. In: The Language Resource and Evaluation Conference (2016)
31. Mihalcea, R.: Graph-based ranking algorithms for sentence extraction, applied to text summarization. In: Proceedings of the ACL 2004 on Interactive poster and demonstration sessions, p. 20. Association for Computational Linguistics (2004)
32. Lin, C.Y.: Rouge: a package for automatic evaluation of summaries. In: Text Summarization Branches Out: Proceedings of the ACL-04 Workshop, vol. 8 (2004)
33. Moro, A., Cecconi, F., Navigli, R.: Multilingual word sense disambiguation and entity linking for everybody. In: Proceedings of ISWC (P&D), pp. 25–28 (2014)
34. Saggion, H.: SUMMA: a robust and adaptable summarization tool. Traitement Automatique des Langues 49(2), 103–125 (2008)
35. Ronzano, F., Fisas, B., Bosque, G.C., Saggion, H.: On the automated generation of scholarly publishing linked datasets: the case of CEUR-WS proceedings. In: Gandon, F., Cabrio, E., Stankovic, M., Zimmermann, A. (eds.) SemWebEval 2015. CCIS, vol. 548, pp. 177–188. Springer, Cham (2015). doi:10.1007/978-3-319-25518-7_15
36. Peroni, S.: The semantic publishing and referencing ontologies. In: Peroni, S. (ed.) Semantic Web Technologies and Legal Scholarly Publishing. Law, Governance and Technology Series, vol. 15, pp. 121–193. Springer, Heidelberg (2014)
37. Thakker, D., Osman, T., Lakin, P.: Gate jape grammar tutorial. Nottingham Trent University, UK, Phil Lakin, UK, Version 1 (2009)
38. Witten, I.H., Frank, E.: Data Mining: Practical Machine Learning Tools and Techniques. Morgan Kaufmann, San Francisco (2005)
39. O'Donoghue, D.P., Abgaz, Y., Hurley, D., Ronzano, F., Saggion, H.: Stimulating and simulating creativity with Dr. Inventor. In: The Proceedings of the International Conference on Computational Creativity (2015)

A Typology of Semantic Relations Dedicated to Scientific Literature Analysis

Kata Gábor[1]([⊠]), Haïfa Zargayouna[1], Isabelle Tellier[2], Davide Buscaldi[1], and Thierry Charnois[1]

[1] LIPN, CNRS (UMR 7030),
Université Paris 13 Sorbonne Paris Cité, Villetaneuse, France
{gabor,haifa.zargayouna,davide.buscaldi,
thierry.charnois}@lipn.univ-paris13.fr
[2] LaTTiCe, CNRS (UMR 8094), ENS Paris,
Université Sorbonne Nouvelle - Paris 3, PSL Research University,
Université Sorbonne Paris Cité, Paris, France
isabelle.tellier@univ-paris3.fr

Abstract. We propose a method for improving access to scientific literature by analyzing the content of research papers beyond citation links and topic tracking. Our model relies on a typology of explicit semantic relations. These relations are instantiated in the abstract/introduction part of the papers and can be identified automatically using textual data and external ontologies. Preliminary results show a promising precision in unsupervised relationship classification.

1 Introduction

Compiling a state of the art is a fundamental activity for the understanding of any scientific research field. This activity requires the analysis of the existing literature to identify the involved concepts and actors and track relevant topics. Chavalarias and Cointet [3], Herrera et al. [8] or Skupin [17] work on analyzing and visualizing the evolution of topics over time. Citation links are extensively used to explore scientific communities [12,13]. [6] provides a list of usual tasks on bibliographies for different classes of users. The Citation Typing ontology (CiTo) [16] presents a typology of citations according to the relation between the research papers they express.

Citations alone, however, are not enough to fully understand the evolution of a research field: researchers need to analyze the contribution of individual papers. Such an analysis is focused on specific concepts and relations, for instance to identify that a *method* has been developed to *tackle* some *problem*, that a *refinement* of an *existing solution* has been *developed*, etc. For this purpose, we need to (i) identify the entities and concepts that describe a scientific field (*method, problem*) and (ii) identify the semantic relations between these entities (*tackle, developed*). Doing so, we will build semantic links between articles that go much beyond explicit citations. Hence, the definition and identification of the relevant semantic relations is at the core of our approach.

© Springer International Publishing AG 2016
A. González-Beltrán et al. (Eds.): SAVE-SD 2016, LNCS 9792, pp. 26–32, 2016.
DOI: 10.1007/978-3-319-53637-8_3

We combine natural language processing techniques with statistical term extractors and external ontological resources. Ontologies allow a fine-tuned semantic analysis, as opposed to Open Information Extraction [5] or general-purpose approaches exploiting terminology extraction on the fly [3,11,12,17]. In particular, systems using an ontology can benefit from various typed relations. As a corpus of scientific texts, we use the ACL Anthology Corpus [15] and we focus on the "abstract" and "introduction" sections, as they provide the most informative description of the content of a paper. However, our approach does not rely on manually annotated data and aims to be domain-independent: the semantic relations considered are generic for any scientific field.

Section 2 introduces the main architecture of our model for semantic analysis. Section 3 proposes a typology of the semantic relations in the scientific domain, while Sect. 4 briefly exposes the methodology by which this model has been instantiated. Finally, Sect. 5 draws some conclusions and perspectives for future work.

2   General Model

Our purpose is to automatically extract relevant semantic relations in the science/engineering domain, as they appear in texts such as "a (new) method is proposed for a task", or "a phenomenon is found in a certain context". By identifying concepts and semantic relations between concepts, we can detect research papers which deal with the same problem, or track the evolution of results on a certain task. Our model of the scientific domain contains scientific articles linked to typed relations whose arguments are mapped to existing ontologies (see Fig. 1).

The process used to implement the model consists of three sub-tasks: entity annotation, concept mapping and relation classification. Entity annotation is the task of recognizing instances of domain concepts in the text. Concept mapping consists in finding mappings with external ontologies or vocabularies. Relation classification uses the entity-annotated text as input and aims to identify the relations between two entities based on a combination of two information sources: the text sequence between the two entities (extracted from the corpus), and the semantic type of the entities (extracted from the ontology). We are currently experimenting with an iterative process: after annotating concepts in the corpus, we extract sequential patterns, which are used to identify instances of known relations and to discover new *types* of relations. A further goal is to enrich ontologies with new relation types [14].

For example, from the text: *"This database contains recorded, transcribed and annotated read speech"* we want to be able to extract the following relation:

```
<relation type="composed_of">
<arg1><entity majorType="BabelNet"synset="bn:00025333n">
database</entity></arg1>
contains (...)
<arg2><entity majorType="BabelNet"synset="bn:00049911n">
speech</entity></arg2></relation>
```

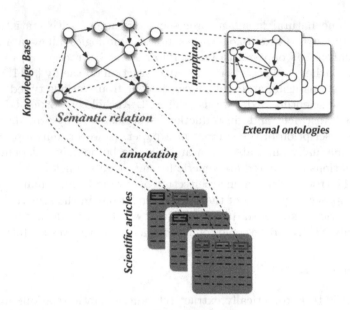

Fig. 1. The general model

3   Semantic Relations in Scientific Literature

A data-driven approach was adopted to discover relation types represented in the corpus. Our corpus contains 4.200.000 words from 11.000 papers (abstracts and introductions) in the ACL Anthology Corpus, pre-processed by E. Omodei [11]. A sample of 100 abstracts was extracted and instances of explicit semantic relations were discovered and manually annotated on these data.

Pattern-based approaches of relationship extraction [1] and classification [18] rely on the hypothesis that the context of occurrence of entity mention pairs is characteristic of the semantic relation between the two concepts. Thus, only linguistically explicit relations were taken into account. On the textual level, a semantic relation is conceived as a text span linking two annotated instances of concepts within the same sentence (see the example above). On the semantic level, relation types need to be specific enough to be easily distinguished from each other by a domain expert. Argument types are very informative when specifying a relation. Instances of arguments are typically domain-specific (e.g. the kind of *data* or *resources* are different across domains), hence, the link is ensured by mapping the entities to external ontologies. Table 1 provides the typology of relations that has been defined, together with argument type specifications. Table 2 shows how the various types of semantic relations are represented in this corpus.

The typology was set up on the basis of examples from the 100 abstracts. As a next step, we selected a sample of 500 abstracts to be manually annotated using the current typology. The agreement rate between the annotators will indicate

Table 1. Semantic relation typology based on 100 abstracts

affects	ARG1: *specific property of data* ARG2: *results*
based_on:	ARG1: *method, system* based on ARG2: *other method*
char	ARG1: *observed characteristics* of an observed ARG2: *entity*
compare	ARG1: *result (of experiment)* compared to ARG2: *result2*
composed_of	ARG1: *database/resource* ARG2: *data*
datasource	ARG1: *information* extracted from ARG2: *kind of data*
method_applied	ARG1: *method* applied to ARG2: *data*
model	ARG1: *abstract representation* of an ARG2: *observed entity*
phenomenon	ARG1: *entity, a phenomenon* found in ARG2: *context*
problem	ARG1: *phenomenon* is a problem in a ARG2: *field/task*
propose	ARG1: *paper/author* presents ARG2: *an idea*
study	ARG1: *analysis* of a ARG2: *phenomenon*
tag	ARG1: *tag/meta-information* associated to an ARG2: *entity*
task_applied	ARG1: *task* performed on ARG2: *data*
used_for	ARG1: *method/system* ARG2: *task*
uses_information	ARG1: *method* relies on ARG2: *information*
yields	ARG1: *experiment/method* ARG2: *result*
wrt	ARG1 *a change* in/with respect to ARG2: *property*

Table 2. Most frequent relations in manually annotated abstracts

Relation	Frequency in corpus
used_for	27%
composed_of	16%
propose	11%
yields	6%
study	6%
task_applied	5%
uses_information	4%
affects	4%

whether the relations are well defined on the semantic level (possible to classify), and whether they are indeed explicit on the textual level (possible to annotate). In case of a successful validation, these data will serve to evaluate relationship extraction and classification experiments.

4   Model Instantiation

Entity annotation was applied to the corpus of 4.2 million words in two steps. First, candidates were generated with the terminology extraction tool TermSuite [4]. The list of extracted terms was then mapped to different ontological resources: the knowledge base of Saffron Knowledge Extraction Framework [2], and the BabelNet ontology [10]. If a term was validated as a domain concept (i.e., found in at least one of the resources), it was annotated in the text. The complete process of entity annotation is described in [7].

A first set of unsupervised, pattern-based clustering experiments was performed to detect semantic relations. First, entity mention pairs were extracted from the corpus, together with the sequences. A co-occurrence matrix was built from entity pairs and sequences. The matrix rows (entity pairs) were clustered using CLUTO's [9] divisive algorithm with repeated bisections. As we are experimenting with completely unsupervised methods, the number of clusters to detect was not fixed to the number of relations in our typology. The reported evaluation was carried out on a set of 700 entity pairs, manually classified to one or more of the semantic relations we defined. Table 3 summarizes the results compared to a random clustering baseline. Precision and recall are calculated in terms of pairs of items correctly or falsely assigned to the same cluster or to different clusters.

Table 3. Evaluation of clustering

Input	#clusters	Precision	Recall	F-measure
Baseline	100	0.095	0.009	0.017
Baseline	50	0.103	0.019	0.033
Baseline	25	**0.104**	**0.041**	**0.058**
Sequences	100	**0.490**	0.046	0.084
Sequences	50	0.378	0.079	0.132
Sequences	25	0.313	**0.140**	**0.193**

5   Conclusion

We presented a model for the analysis of scientific papers in order to automatically extract states of the art of a research field. The core of the model is a typology of semantic relations in the scientific domain, which was defined while manually annotating data from a corpus of natural language processing papers. These relations can be identified automatically using a combination of pattern mining and natural language processing techniques. The first results on recognizing relations between unseen concepts are already very encouraging: a precision of 0.5 means that one out of two pairs of concepts assigned to the same category belong to the same semantic relation (among 18 distinct possible ones). Experiments are currently being carried out with bi-clustering algorithms, where text sequences and concept pairs are clustered at the same time.

Acknowledgments. This work is part of the program "Investissements d'Avenir" overseen by the French National Research Agency, ANR-10-LABX-0083 (Labex EFL).

References

1. Auger, A., Barrière, C.: Pattern-based approaches to semantic relation extraction: a state-of-the-art. Terminology **14**(1), 1–19 (2008)
2. Bordea, G., Buitelaar, P., Polajnar, T.: Domain-independent term extraction through domain modelling. In: 10th TIA Conference (2013)
3. Chavalarias, D., Cointet, J.-P.: Phylomemetic patterns in science evolution - the rise and fall of scientific fields. PLoS ONE **8**(2) (2013)
4. Daille, B., Jacquin, C., Monceaux, L., Morin, E., Rocheteau, J.: TTC TermSuite: Une chaîne de traitement pour la fouille terminologique multilingue. In: Proceedings of the TALN Conference (2013)
5. Del Corro, L., Gemulla, R.: ClausIE: clause-based open information extraction. In: Proceedings of the 22nd WWW Conference, pp. 355–366. International World Wide Web Conferences Steering Committee, Republic and Canton of Geneva (2013)
6. Di Iorio, A., Giannella, R., Poggi, F., Vitali, F.: Exploring bibliographies for research-related tasks. In: Proceedings of the 24th International Conference on World Wide Web Companion, pp. 1001–1006. International World Wide Web Conferences Steering Committee (2015)
7. Gabor, K., Zargayouna, H., Buscaldi, D., Tellier, I., Charnois, T.: Semantic annotation of the ACL anthology corpus for the automatic analysis of scientific literature. In: Proceedings of the LREC 2016 Conference, Portoroz, Slovenia (2016)
8. Herrera, M., Roberts, D.C., Gulbahce, N.: Mapping the evolution of scientific fields. PLoS ONE **5** (2010)
9. Karypis, G.: CLUTO: A clustering toolkit. Technical report 02–017, Department of Computer Science, University of Minnesota (2002)
10. Navigli, R., Ponzetto, S.P.: BabelNet: the automatic construction, evaluation and application of a wide-coverage multilingual semantic network. Artif. Intell. **193**, 217–250 (2012)
11. Omodei, E., Cointet, J.-P., Poibeau, T.: Mapping the natural language processing domain: experiments using the ACL anthology. In: Proceedings of the LREC 2014 Conference (2014)
12. Osborne, F., Motta, E.: Mining semantic relations between research areas. In: International Semantic Web Conference, Boston, MA (2012)
13. Osborne, F., Motta, E.: Rexplore: unveiling the dynamics of scholarly data. Digit. Libr. **8**(12) (2014)
14. Petasis, G., Karkaletsis, V., Paliouras, G., Krithara, A., Zavitsanos, E.: Ontology population and enrichment: state of the art. In: Paliouras, G., Spyropoulos, C.D., Tsatsaronis, G. (eds.) Knowledge-Driven Multimedia Information Extraction and Ontology Evolution. LNCS (LNAI), vol. 6050, pp. 134–166. Springer, Heidelberg (2011). doi:10.1007/978-3-642-20795-2_6
15. Radev, D.R., Muthukrishnan, P., Qazvinian, V.: The ACL anthology network corpus. In: Proceedings of the 2009 ACL Workshop on Text and Citation Analysis for Scholarly Digital Libraries (2009)

16. Shotton, D.: CiTO, the citation typing ontology. J. Biomed. Seman. **1**(S–1), S6 (2010)
17. Skupin, A.: The world of geography: visualizing a knowledge domain with cartographic means. Proc. Nat. Acad. Sci. **101**, 5274–5278 (2004)
18. Turney, P.D.: Similarity of semantic relations. Comput. Linguist. **32**(3), 379–416 (2006)

Text Mining of Related Events
from Natural Science Literature

Erwin Marsi$^{(\boxtimes)}$ and Pinar Özturk

Department of Computer and Information Science,
Norwegian University of Science and Technology, Trondheim, Norway
{emarsi,pinar}@idi.ntnu.no

Abstract. We present an approach to text mining in areas where
the entities of interest can not be defined in advance. Our system is
aimed at finding related events in natural science literature, in particular, changing/increasing/decreasing variables in Marine science publications. It enables semantic search for events by abstracting from
morphological, lexical-semantic and syntactic variations. In addition,
generalisation of variables through syntactic pruning helps finding
similar variables. Relations between events are induced from co-
occurrence frequencies. Extracted information is stored in a property
graph database and accessed using the Cypher query language. A user
interface presents events as a graph to visualise their type, frequency and
relation strength, in combination with their textual sources.

Keywords: Text mining · Natural language processing · Information
extraction · Visualisation · Knowledge discovery

1   Introduction

Text mining of scientific literature originates from efforts to cope with the ever
growing flood of publications in biomedicine [1]. Consequently the resulting
approaches, methods, resources and applications are rooted in the paradigm
of biomedical research and its conceptual framework [2]. Text mining is now
finding its way to other scientific disciplines, promising support for knowledge
discovery from large text collections. Our own research targets text mining in
marine science. As text mining efforts in this area are extremely rare [5–7], it is
not surprising that a corresponding infrastructure is mostly lacking. Moreover,
we found that due to significant differences between the conceptual frameworks
of biomedicine and marine science, simply "porting" the biomedical text mining
infrastructure will not suffice. One major difference is that the biomedical entities
of interest are relatively well defined – genes, proteins, organisms, species, drugs,
diseases, etc. – and typically expressed as proper nouns. In contrast, defining the
entities of interest in marine science turns out to be much harder. Not only does
it seem to be more open-ended in nature, the "entities" themselves tend to be
complex and expressed as noun phrases containing multiple modifiers, giving

© Springer International Publishing AG 2016
A. González-Beltrán et al. (Eds.): SAVE-SD 2016, LNCS 9792, pp. 33–38, 2016.
DOI: 10.1007/978-3-319-53637-8_4

rise to examples like *timing and magnitude of surface temperature evolution in the Southern Hemisphere in deglacial proxy records.*

Theories and models in marine science typically involve changing variables and their complex interactions, which includes correlations, causal relations and chains of positive/negative feedback loops. Many marine scientists are interested in finding evidence – or counter-evidence – in the literature for events of change and their relations. Given the difficulties with defining entities, we focus on mining of these events, leaving entities underspecified for the time being, simply referring to them as *variables*. Here we describe ongoing work to automatically extract, relate, query and visualise events of change and their direction of variation: *increasing*, *decreasing* or just *changing* (i.e. direction not specified in the text).

2   Approach

Our system is essentially a pipeline involving a number of processing steps.

Information retrieval – The first step is collecting publications of interest – for our use case, Marine science articles concerning the biological pump and/or food webs. Our text material consists of abstracts from selected journals by Nature Publishing Group. Search terms obtained from domain experts were used to query Nature's OpenSearch API[1] for publications in selected journals, after 1997, retrieving records including title and abstract. The top-10k abstracts matching most search terms were selected for further processing with the Stanford CoreNLP tools [4], including tokenisation, sentence splitting, part-of-speech tagging, lemmatisation and syntactic parsing. Lemmatised parse trees were obtained by substituting terminals with their lemmas. The resulting new corpus contains 9,884 article abstracts, 29,565 sentences and approximately 626 k tokens.

Information extraction – The second step extracts change events and the variables they pertain to. Tree pattern matching is applied to lemmatised syntax trees using the Tregex engine [3], which provides a compact language for writing regular expressions over trees. Seven hand-written pattern templates were instantiated with lexical instances from manually created lists of verbs and nouns expressing change, yielding 320 tree matching patterns. The total number of matched variables in the corpus is 22,784: 9,673 for change, 7,827 for increase and 5,289 for decrease. For more details, see [5].

Generalisation of variables – Since many of the extracted variables are long and complex expressions, their frequency is low. The most frequent variables are generic terms (*climate* 1350, *temperature* 165, *global climate* 86), but over 66% is unique. This evidently impedes the discovery of relations among events. As a partial solution to this problem, variables are generalised by progressive pruning of syntax trees using a set of tree transformation operations. This effectively produces more abstract variants of variables. For example, the variable

---

[1] http://www.nature.com/developers/documentation/api-references/opensearch-api.

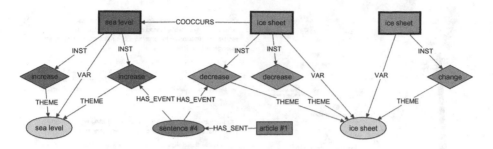

Fig. 1. Event model in the graph database (Color figure online)

the annual, Milankovitch and continuum temperature is split into three variables
– *annual temperature, Milankovitch temperature* and *continuum temperature* –
all which are ultimately reduced to just *temperature*. Tree transformations are
implemented using Tsurgeon [3]: Tregex patterns match the syntactic structures
of interest, whereas an associated Tsurgeon operation deletes selected nodes (cf.
[5]) Generalisation resulted in 102,625 variables, which is 4.5 times the number
of originally extracted variables.

Graph creation – The extracted events are stored in a property graph database
as nodes, directed edges and associated properties. Figure 1 shows a small partial
sample of how events are modelled. The diamond-shaped nodes represent events,
with red for an INCREASE, blue for a DECREASE and green for a CHANGE events.
An event pertains to a unique VARIABLE type (yellow nodes) as indicated by
its THEME edge. Each event also occurs in a SENTENCE through a HAS-EVENT
edge, which in turn is linked to an ARTICLE via a HAS-SENT edge. Aggregated
nodes join all event nodes with the same type and variable. For example, the
square blue node labelled "ice sheet" joins all event instances where the variable
"ice sheet" is decreasing. Likewise, the square red node labelled "sea level" joins
all sea level decrease events. Finally, aggregated nodes can be connected by
COOCCURS edges whenever two events co-occur in a single sentence. For example,
a decrease of "ice sheet" co-occurs with an increase of "sea level" in sentence
number 4. In addition, nodes and edges have properties which hold important
information. For instance, SENTENCE nodes hold the sentence string, event nodes
hold the character offsets for their variable string and COOCCURS edges hold the
frequency of co-occurrence. The Neo4j graph database[2] (community edition) is
used for storing and accessing the graph. The powerful Cypher query language
– akin to SQL for relational databases – makes it relatively easy to search for
relations between events, for example, to find the shortest path between an
increase in A and a decrease in B over any number of co-occurrence links.

---

[2] http://neo4j.com/.

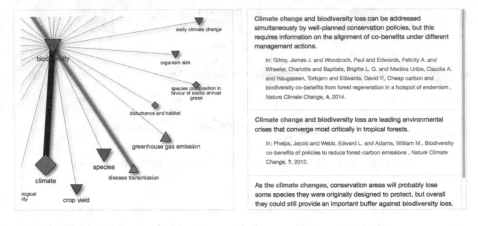

Fig. 2. Visualisation of event relations (Color figure online)

User interface – These search capabilities are partly exposed to end users through a web application with a graphical user interface. Users can search for single events, relations between pairs of events or even triples of related events (i.e. indirect relations). Each event can optionally be restricted by type (increase, decrease or change) and by variable involved. The type of relation between events is currently limited to co-occurrence, but will be extended to causal relations and correlations in the near future. Figure 2 shows part of the output for any events related to a *decrease in biodiversity*. The left pane shows a graph where the nodes are events – red for increase, blue for decrease and green for change – labelled with their variable.[3] The node size corresponds to an event's overall frequency, whereas edge weight denotes co-occurrence frequency. The graph can be moved/resized and otherwise filtered and formatted according to a user's need. Selecting an edge (black edge in Fig. 2) will list the corresponding source sentences in the right pane, with highlighted variables and links to original article web pages.

3   Discussion

We have presented an approach to text mining from natural science literature which is aimed at finding related events. It provides semantic search for events in the sense that it abstracts from morphological variations (e.g. singular/plural), lexical-semantic variations (e.g. an increase can be expressed by *rise, enhance, boost*, etc.), syntactic variations (e.g. *X increases, something increases X, increasing X, X is increasing, an X increase, increase in X*, etc.). In addition, generalisation of variables through syntactic pruning helps finding similar variables: for example, both *the annual, Milankovitch and continuum temperature variability* and *annual temperature between 1958 and 2010* are progressively generalised to

---

[3] Graph rendering with vis.js Javascript library: http://visjs.org/.

annual temperature, revealing their similarity at a more abstract level. Whereas a more elaborate description of the information retrieval, extraction and generalisation steps was presented in [5], novel contributions here include the graph model and the user interface. Events, variables and other information are stored in a property graph database and can thus be easily accessed, traversed or modified using the Cypher query language. A user interface presents events as a graph to visualise their type, frequency and relation strength, also providing links their textual sources. We believe the approach is general and applicable to other areas where the entities of interest can not be defined in advance (with minor adaptations of patterns and lexical items).

The current implementation is a proof of concept, but produces a fair amount of noise. Analysis suggests that most problems originate from syntactic parsing errors (in particular coordination and prepositional phrase attachment). As a result, patterns may either fail to match or match unintentionally, yielding incomplete or incoherent variables. Pruning variables is beneficial but insufficient and should be supplemented with other methods. For instance, linking named entities like species, chemicals or geographical locations to unique concepts in appropriate ontologies/taxonomies would allow for generalisations such as *iron* is a *metal* or a *diatom* is a *plankton*. Likewise co-occurrence frequency is a weak signal and part of our ongoing work is therefore to extract causal relations and correlations between events using both pattern matching and machine learning methods. Ultimately events obtained from different publications can be chained together, often with the help of domain knowledge, in order to generate new hypotheses, as pioneered in the work on literature-based knowledge discovery [8]. We will release the source code of a more mature version of our software, as well as various data sets of extracted events, in the near future.

Acknowledgements. Financial aid from the European Commission (OCEAN-CERTAIN, FP7-ENV-2013-6.1-1; no: 603773) is gratefully acknowledged. We thank Murat Van Ardelan for sharing his knowledge of Marine science.

References

1. Ananiadou, S., Mcnaught, J.: Text Mining for Biology And Biomedicine. Artech House Inc., Norwood (2005)
2. Cohen, K.B., Hunter, L.: Getting started in text mining. PLoS Comput. Biol. **4**(1), e20 (2008)
3. Levy, R., Andrew, G.: Tregex and Tsurgeon: tools for querying and manipulating tree data structures. In: Proceedings of ELREC, pp. 2231–2234 (2006)
4. Manning, C.D., Surdeanu, M., Bauer, J., Finkel, J., Bethard, S.J., McClosky, D.: The stanford CoreNLP natural language processing toolkit. In: Proceedings of ACL, pp. 55–60 (2014)
5. Marsi, E., Öztürk, P.: Extraction and generalisation of variables from scientific publications. In: Proceedings of EMNLP, pp. 505–511, Lisbon, Portugal (2015)
6. Marsi, E., Öztürk, P., Aamot, E., Sizov, G., Ardelan, M.V.: Towards text mining in climate science: extraction of quantitative variables and their relations. In: Proceedings of Fourth Workshop on Building and Evaluating Resources for Health and Biomedical Text Processing, Reykjavik, Iceland (2014)

7. Radom, M., Rybarczyk, A., Kottmann, R., Formanowicz, P., Szachniuk, M., Glöckner, F.O., Rebholz-Schuhmann, D., Błażewicz, J.: Poseidon: an information retrieval and extraction system for metagenomic marine science. Ecol. Inf. **12**, 10–15 (2012)
8. Swanson, D.R.: Fish oil, Raynaud's syndrome, and undiscovered public knowledge. Perspect. Biol. Med. **30**(1), 7–18 (1986)

From Papers to Triples: An Open Source Workflow for Semantic Publishing Experiments

Bahar Sateli[✉] and René Witte

Semantic Software Lab, Department of Computer Science
and Software Engineering, Concordia University, Montréal, Canada
sateli@semanticsoftware.info

Abstract. In this demonstration paper, we describe an open source workflow for supporting experiments in semantic publishing research. Based on a flexible, component-based approach, natural language papers can be converted into a Linked Open Data (LOD) compliant knowledge base. We exemplary discuss how to plan and execute experiments based on an integrated suite of tools, thereby both significantly lowering the barrier of entry in this field, while also encouraging the exchange of tools when building novel contributions.

1 Introduction

Semantic publishing research aims at making scientific publications readable and semantically understandable to computers. The long-term vision is to enable automated agents supporting researchers in their daily work: Finding literature pertaining to a task, automatically summarizing state-of-the-art, connecting experiments and datasets, or detecting novel contributions in a field. A number of approaches in this area build on the standards and tools from the Semantic Web initiative [1], such as the Resource Description Framework (RDF) and its vocabularies, RDF Schema (RDFS), which are by now well-supported through numerous open source tools.

When dealing with the huge amount of existing literature, a required foundation for performing experiments in this area is an automated, robust process for converting natural language texts into a Linked Open Data (LOD) [4] compliant format. The resulting knowledge base can then be easily inter-linked with other (research) entities on the web of data, supporting numerous use cases in semantic publishing research. Here, we describe a corresponding workflow and its implementation, based on a combination of our own with existing open source infrastructure, including natural language processing (NLP) components, entity grounding to the LOD cloud, and converting NLP results to RDF triples. This approach has been successfully applied in a number of semantic publishing experiments, including building semantic wiki user interfaces for literature management [7], semantic literature querying [9], the Semantic Publishing Challenge (2015) Task 2 [8], and semantic user profiling of scientists based on their publications [6].

© Springer International Publishing AG 2016
A. González-Beltrán et al. (Eds.): SAVE-SD 2016, LNCS 9792, pp. 39–44, 2016.
DOI: 10.1007/978-3-319-53637-8_5

2   Converting Papers to Triples: The Processing Workflow

The approach demoed here has the core assumption that all relevant information for building semantic publishing applications is extracted from textual artifacts and inserted into a LOD-compliant knowledge base (KB). We start from a set of documents, typically scientific articles (e.g., conference papers or journal articles), but possibly also other texts, like dataset descriptions or scientific tool documentations. For the scope of this demonstration, we exclude a discussion on text extraction from PDF documents; a comprehensive overview can be found in [2]. By default, the GATE text analysis framework [3] used here relies on the Apache Tika library[1] for converting different file types, such as Word, ODT, PPT, or PDF, to plain text.

Semantic analysis tools running on the textual documents provide structured descriptions, for example, on entities, citations, rhetorical entities, or writing styles. To make these results available in a form of a knowledge base, we show how we can directly connect the GATE framework with a triplestore using a novel approach we developed based on Apache Jena,[2] the *LODeXporter*. The resulting KB can then be queried for further scientific experiments or leveraged within a user interface, as illustrated in Fig. 1.

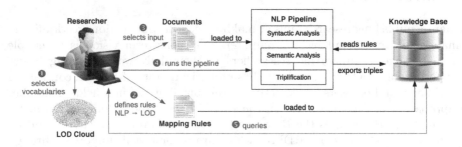

Fig. 1. Workflow for the triplification of scientific literature

2.1   Text Mining Components

The natural language analysis part of our workflow is implemented based on the GATE *(General Architecture for Text Engineering)* framework [3], which provides us with a robust and widely used NLP infrastructure. GATE is designed as a component-based architecture, where individual analysis components (called *processing resources* or PRs) can be easily added, modified, or removed from a system. A document is processed by a sequential *pipeline* of PRs: Each component can read and add results to a text in form of *annotations*, which form a graph over the document. GATE is licensed under the GNU LGPL and can be

---

[1] Apache Tika, https://tika.apache.org/.
[2] Apache Jena, https://jena.apache.org/.

obtained from http://gate.ac.uk; GATE Embedded libraries are also available through Maven Central. Most of the plugins described below can be easily added through GATE's *Plugin Manager* from our *Semantic Software Lab* repository. Snapshots of our presented code are additionally available on our GitHub repository at https://github.com/SemanticSoftwareLab. Continuous integration services are provided through a Jenkins server at http://assistant.semanticsoftware. info/.

Preprocessing. For preprocessing documents, we rely on the standard components shipped with the GATE distribution, in particular the ANNIE plugin [3]. These perform standard NLP tasks that later steps build upon, such as tokenization, sentence splitting, part-of-speech (POS) tagging, stemming, and verb group analysis.

Rhetector. A distinguishing feature of scientific literature, compared to other textual documents, is that sections of a scholarly document usually follow a specific argumentative order. In fact, several de-facto standards exist, such as IMRAD [10], to capture the authors' rhetoric in various domains, with the aim of making scientific communication efficient and organized. A challenging task in the process of text mining scientific literature is to automatically detect these rhetoric zones in a document. The automatic semantic annotation of Rhetorical Entities (REs), such as *contributions* or *claims*, has proven to be effective in finding information on a more granular level, for example, in finding all papers that use a specific method M in their experiments [9]. Here, we show our *Rhetector* component[3] to automatically detect REs in scientific literature, currently limited to Claims and Contributions. For each detected RE, an annotation of type "RhetoricalEntity" is added to the document. Based on the grammatical structure of the detected RE, it is then classified and mapped onto existing concepts on the Linked Open Data (LOD) cloud. Figure 2 shows a number of hand-crafted rules written in GATE's JAPE language [3] that incrementally detect *deictic* and

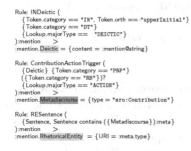

(a) Example JAPE rules

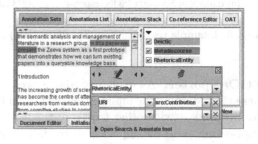

(b) Detected RE annotation in GATE Developer

Fig. 2. JAPE rules (left) to extract a **Contribution** sentence and the generated annotations, color-coded in GATE's Developer GUI

---

[3] Rhetector, http://www.semanticsoftware.info/rhetector.

meta-discourse entities in text and classify the encompassing sentence, based on the discourse actions, as a new rhetorical entity. Our Rhetector PR is licensed under the GNU LGPL v3 and available through GATE's Plugin Manager.

LODtagger. To detect domain-specific entities in research publications, we rely on the LOD cloud, in particular DBpedia. This provides for a rich, continuously updated resource in a standard semantic format. By linking entities detected in documents to LOD URIs (Universal Resource Identifiers), we can semantically query a knowledge base for all papers on a specific topic (URI), even when that topic is not mentioned literally in a text: E.g., we can find a paper for the topic *"linked open data,"* even when it only mentions *"LOD,"* since they are semantically related in the DBpedia ontology. For the actual entity tagging, we rely on an external tool, DBpedia Spotlight [5]. To integrate this web service into a GATE text mining pipeline, we developed *LODtagger*.[4] This component sends the entire UTF-8 formatted text of a document as a RESTful POST request to a Spotlight endpoint and receives the results in JSON format, which are subsequently parsed and transformed to GATE annotations (Fig. 3). To further limit the annotations to named entities, we also demonstrate how we can filter LODtagger results using our syntactic MuNPEx[5] noun phrase (NP) chunker, thereby significantly reducing false positives. Our LODtagger PR is also licensed under the GNU LGPL v3 and can be installed through GATE's Plugin Manager.

(a) Excerpt of Spotlight JSON response   (b) Generated NE annotation in GATE

Fig. 3. Example response from Spotlight (left) and the generated annotation (right)

LODeXporter. As explained above, we aim to create a semantic knowledge base that contains the information extracted from research documents, to be able to inter-link it with other triples in the same KB or the LOD cloud, ultimately supporting end-user applications or data analysis experiments. However, so far we only generated GATE-specific document annotations. Rather than 'hard-coding' a specific export strategy of GATE annotations to triples, we wanted a more flexible solution that can *(a)* be easily extended for new NLP annotations (e.g., when importing new PRs into a pipeline) and *(b)* provide flexible mapping

---

[4] LODtagger, http://www.semanticsoftware.info/lodtagger.

[5] MuNPEx, http://www.semanticsoftware.info/munpex.

of annotations to LOD vocabularies, in order to facilitate experiments with different ontologies. Our solution is a novel process where we derive the export process, including the mapping of NLP entities to LOD vocabularies, from a knowledge base. Our *LODeXporter* component directly connects a GATE pipeline to a (currently Jena TDB-based) triplestore. The KB contains rules, expressed in RDF, which describe how a specific GATE annotation should be mapped to triples; as well as the vocabularies to use, such as the one shown below that describes the mapping of a GATE "DBpediaNE" annotation to an RDF triple of type "pubo:LinkedNamedEntity":

```
@prefix map: <http://semanticsoftware.info/mapping/mapping#> .
@prefix pubo: <http://lod.semanticsoftware.info/pubo/pubo#> .

map:GATEDBpediaNE a map:Mapping ;
    map:GATEtype    "DBpediaNE" ;
    map:type        pubo:LinkedNamedEntity ;
```

This way, the same pipeline can support different RDF export results, simply by virtue of changing the triples in the KB or connecting to a different KB. LODeXporter[6] is currently considered to be in pre-release, as we plan to improve the expressiveness of mapping relations between entities before the 1.0 release.

2.2 Application

At this point, we have a knowledge base populated with the information extracted from research publications. We can now leverage this KB for experiments, by querying it for specific information using SPARQL, e.g., by deploying an Apache Fuseki[7] server. The general challenge here is to formulate a scientific hypothesis that can be empirically evaluated based on the query results. In this demo, we show how we executed and evaluated a number of concrete tasks, such as querying documents based on entities alone vs. entities appearing in rhetorical zones [9] or matching documents to semantic user profiles [6]. While some experiments can be performed solely based on the generated triples, others will require a user evaluation or comparison to a gold standard. However, we cannot cover these steps within the scope of this demonstration paper.

Of course, the generated KB can also be used to drive end-user applications for evaluating the impact of semantic support on concrete scholarly tasks, such as literature review – for example, through a semantic wiki system like *Zeeva* [7].

3 Conclusions

We described a workflow for experiments in semantic publishing research that is based entirely on open source tools and standards. The demo will highlight how this process can be easily configured for different research questions. We hope that the presented work will help others to adapt semantic text mining tools in their projects.

---

[6] LODeXporter, http://www.semanticsoftware.info/lodexporter.

[7] Apache Fuseki, https://jena.apache.org/documentation/serving_data/.

References

1. Berners-Lee, T., Hendler, J.: Publishing on the semantic web. Nature 410(6832) (2001)
2. Constantin, A., Pettifer, S., Voronkov, A.: PDFX: fully-automated PDF-to-XML conversion of scientific literature. In: Proceedings of the 2013 ACM Symposium on Document Engineering (DocEng 2013), pp. 177–180. ACM, New York (2013)
3. Cunningham, H., Maynard, D., Bontcheva, K., Tablan, V., Aswani, N., Roberts, I., et al.: Text Processing with GATE (Version 6) (2011). http://tinyurl.com/gatebook
4. Heath, T., Bizer, C.: Linked Data: Evolving the Web into a Global Data Space. Synthesis lectures on the semantic web: theory and technology. Morgan & Claypool Publishers (2011)
5. Mendes, P.N., Jakob, M., García-Silva, A., Bizer, C.: DBpedia spotlight: shedding light on the web of documents. In: Proceedings of the 7th International Conference on Semantic Systems, pp. 1–8. ACM (2011)
6. Sateli, B., Löffler, F., König-Ries, B., Witte, R.: Semantic user profiles: learning scholars' competences by analyzing their publications. In: Semantics, Analytics, Visualisation: Enhancing Scholarly Data (SAVE-SD 2016), Montreal, QC, Canada, 11 April 2016
7. Sateli, B., Witte, R.: Supporting researchers with a semantic literature management wiki. In: The 4th Workshop on Semantic Publishing (SePublica 2014). CEUR Workshop Proceedings, vol. 1155. Anissaras, Crete, Greece, 25 May 2014. http://ceur-ws.org/Vol-1155/paper-03.pdf
8. Sateli, B., Witte, R.: Automatic construction of a semantic knowledge base from CEUR workshop proceedings. In: Gandon, F., Cabrio, E., Stankovic, M., Zimmermann, A. (eds.) SemWebEval 2015. CCIS, vol. 548, pp. 129–141. Springer, Cham (2015). doi:10.1007/978-3-319-25518-7_11
9. Sateli, B., Witte, R.: Semantic representation of scientific literature: bringing claims, contributions and named entities onto the Linked Open Data cloud. PeerJ Comput. Sci. 1(e37) (2015). https://peerj.com/articles/cs-37/
10. Sollaci, L.B., Pereira, M.G.: The introduction, methods, results, and discussion (IMRAD) structure: a fifty-year survey. J. Med. Libr. Assoc. 92(3), 364 (2004)

OpenAIRE LOD Services: Scholarly Communication Data as Linked Data

Giorgos Alexiou[1,2], Sahar Vahdati[3(✉)], Christoph Lange[3,4],
George Papastefanatos[1], and Steffen Lohmann[3,4]

[1] Institute for the Management of Information Systems,
Athena Research Center, Athens, Greece
{galexiou,gpapas}@imis.athena-innovation.gr
[2] School of Electrical and Computer Enginnering,
National Technical University of Athens, Athens, Greece
[3] Enterprise Information Systems (EIS), University of Bonn, Bonn, Germany
vahdati@uni-bonn.de, langec@cs.uni-bonn.de,
steffen.lohmann@iais.fraunhofer.de
[4] Fraunhofer Institute for Intelligent Analysis and Information Systems (IAIS),
Sankt Augustin, Germany

Abstract. OpenAIRE, the Open Access Infrastructure for Research in Europe, enables search, discovery and monitoring of publications and datasets from more than 100,000 research projects. Increasing the reusability of the OpenAIRE research metadata, connecting it to other open data about projects, publications, people and organizations, and reaching out to further related domains requires better technical interoperability, which we aim at achieving by exposing the OpenAIRE Information Space as *Linked Data*. We present a scalable and maintainable architecture that converts the OpenAIRE data from its original HBase NoSQL source to RDF. We furthermore explore how this novel integration of data about research can facilitate scholarly communication.

1 Introduction

OpenAIRE $(OA)^1$ is the European Union's flagship project for an Open Access Infrastructure for Research; it enables search, discovery and monitoring of scientific outputs (more than 13M publications, 12M authors and scientific datasets), harvested from over 6 K data providers and linked to more than 100 K research projects funded by EU and Australian bodies. To increase the interoperability of the OA Information Space (IS), we have published its data as Linked Open Data (LOD). In our previous work [6], we have specified a vocabulary for the OA LOD and experimented with different implementations of publishing the OA IS as LOD. Based on this preliminary work, we have developed and now present a scalable implementation over Hadoop that can efficiently address the publishing of large volumes of scholarly data, through which OA can offer three different

---

1 http://www.openaire.eu.

© Springer International Publishing AG 2016
A. González-Beltrán et al. (Eds.): SAVE-SD 2016, LNCS 9792, pp. 45–50, 2016.
DOI: 10.1007/978-3-319-53637-8_6

LOD services: i. fine-grained exploration of data records about individual entities in the OA IS, ii. a downloadable all-in-one data dump, and iii. interactive
querying via a SPARQL endpoint, i.e., a standardized query interface. On top
of this setup we can add further services, e.g., for visual exploration or data
analysis, and proceed with linking the OA data to related datasets.

The OA infrastructure is a data *aggregator* rather than a primary producer,
i.e., it processes information from many different repositories in arbitrary harvesting cycles. In this setting, the process of publishing and interlinking scholarly
data as LOD has revealed a number of interesting technical challenges. A typical
problem is related to the *persistent identification* of published entities. Harvesting information from multiple, inherently dynamic and heterogeneous sources
leads to duplication of content; thus, deduplication before publishing aggregated
data is a common practice. Deduplication identifies groups of entities that represent the same real-world object (e.g. author) based on schema and content
characteristics [4] and merges them into one representative record. Content harvesting and deduplication are repeatedly performed to sync the IS with updated
information at the sources; however, these processes do not guarantee persistent
identifiers for the disambiguated entities. Thus, we enhanced the OA Data Model
by temporal characteristics to ease tracking changes between updates in the IS.
A second challenge relates to the *performance* of the LOD production process
and its *scalability* to the huge data volume. The process must be performed efficiently, such that it seamlessly integrates into the OA data lifecycle, avoiding
the provisioning of outdated LOD. Therefore, we pursue a parallel Map-Reduce
processing strategy.

Our first step was to model the Linked Data vocabulary (ontology) and the
mappings between the OA Data model entities and the ontology classes. For the
publishing process, we convert and assign URIs to all individual records except
representative ones. This is performed incrementally, using temporal annotations, such that only new or updated records are converted. The result is stored
in an RDF triple store. Next, we process and store all information concerning
duplicate relations (e.g. *sameAs*) between the aforementioned records. The reason for excluding *representative* records is that their identification is based on the
duplicate records they are derived from, which, given the evolving nature of the
sources and the varying performance of deduplication, is not persistent across
harvesting cycles (even if the original entities stay intact). Instead, we choose to
publish all the original records and explicitly mark them as duplicates with the
`owl:sameAs` property. Our approach has been implemented as a Hadoop workflow, and integrated into the OA production system as a parallel job to all other
data processing activities.

2   OpenAIRE LOD Framework

The OA LOD framework aims at providing a set of services for publishing OA
resources as LOD and providing an infrastructure for data access, retrieval and
citation (e.g., a SPARQL endpoint or a LOD API); Furthermore, one of its main

purposes is interlinking with popular LOD datasets and services (DBLP, ACM, CiteSeer, DBpedia, etc.) and enriching the OA IS with information from the LOD cloud. The OA LOD is downloadable as a dump through http://lod. openaire.eu[2] and queryable via a SPARQL endpoint. According to the recommended best practices [2], we use content negotiation to handle incoming HTTP requests: requests from Linked Data clients, which ask for an RDF-specific media type (i.e., `application/rdfxml+`) in their HTTP header, are answered by the RDF store, while all other HTTP requests to http://lod.openaire.eu are answered with human-readable HTML pages.

2.1 OpenAIRE Linked Data Vocabulary

An major requirement for designing the OA LOD vocabulary was to reuse concepts, properties and terms from existing standards and initiatives, to maximize the interoperability of the OA LOD with other data sources. Given the rich OA data model, the main challenges were to identify the most suitable vocabularies for reuse, but also to define our own, i.e., OA specific vocabulary terms for attributes not captured by existing vocabularies. As the schema of the OA LOD, we specified an OWL ontology by mapping the entities of the OA data model to OWL classes, and its attributes and relationships to OWL properties. Vocabularies reused include Dublin Core for general metadata, SKOS for classification and CERIF[3] for research organizations and activities. We linked new, OA-specific terms to reused ones, e.g., by declaring `Result` a superclass of http://purl.org/ontology/bibo/AcademicArticle and http://www.w3.org/ns/ dcat#Dataset.

For the URI scheme, our goal was to assign user-friendly URIs; though this was partially impossible because of inherent restrictions of OA's current way of identifying entities. As *base URI*, we use our own domain with the `data` path to distinguish actual resources from pages about the resources, i.e., http://lod.OpenAIRE.eu/data/. Subsequently, we add the type of each resource (Datasource, Organization, Person, Project and Result) represented by a URI, and finally add the unique identification of that resource, i.e., http://lod.openaire.eu/data/organization/{id}.

2.2 LOD Production

In the following, we present the technical details of our framework. The data of the OA IS is available in three source formats: HBase (a NoSQL database), XML and CSV. A comparison of mappings from each of these three source formats to RDF led to the observation that mapping from HBase may be faster in terms of performance, however, mapping from CSV is not significantly slower but at the same time much more maintainable; it is thus our preferred option [6].

---

[2] For the moment, this URL redirects to http://beta.lod.openaire.eu to indicate that the OpenAIRE LOD Services are currently in beta.

[3] Common European Research Information Format (http://www.eurocris.org/cerif/ main-features-cerif.).

The first mapping step involves the RDFization process. This process takes as input **two** CSV files, one with all records, and a second one with all the relations about duplicate records, converts them to RDF and stores them as separate named graphs in our RDF triple store. The first graph, which holds all OA entities, is the largest graph and is updated incrementally based on **temporal** properties that we have introduced in both the OA vocabulary and data model while the second graph, which holds all the relationships, is a small graph that is dropped and recreated in every run of our workflow following the output of the deduplication process.

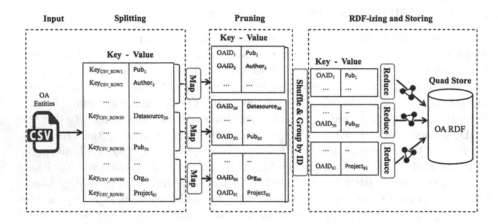

Fig. 1. RDFization of the OpenAIRE IS with Hadoop parallel processing.

Figure 1 shows the functionality of our approach in terms of M/R jobs. In the first step, the CSV file that contains the entities is loaded (*input*) and automatically split (*splitting*) by the Hadoop framework into smaller chunks and distributed between the mappers. The entities are split in *key-value pairs*, where the *key* is an ID auto-assigned by the framework and the *value* is the actual entity. Then, mappers parse the CSV and map the entities from the CSV according to *map(ID, value)* where *ID=OAIDID* and *value=entity_attributes*. In that stage, we omit entities whose last modification date precedes the last execution date of the process (*pruning*). The output is usually a small subset of our initial input, containing only the entities that have changed since the last execution of our workflow.

Subsequently, the output is *shuffled and grouped by ID* and distributed to the *reducers*. *ID=OAIDID* was selected as key because Hadoop's default hashcode-based partitioning algorithm distributes entities uniformly to reducers based on their IDs (`ID.hashCode() mod numReduceTasks`). Finally, the *reducers* extract each input entity's attributes, convert them to RDF and store them directly in the RDF store (*RDFizing and storing*). We insert the data directly to the RDF store instead of saving it to HDFS and then loading it to our database. With the use of appropriate connection pooling, our RDF store (OpenLink Virtuoso) can scale and handle efficiently the aforementioned approach largely automatically.

3   Related Work

OAI2LOD Server [2] is a tool designed to publish Linked Data content from aggregators and repositories which are compatible with the Open Archives Initiative Protocol for Metadata Harvesting (OAI-PMH). While it follows most of the Linked Data directives concerning URI design, it currently does not sufficiently address the problem of URI persistence and data volume, in terms of scalability [3].

Among the first adopters of the Linked Data approach in the digital libraries community was the Library of Congress (LoC) [5]. It exposes millions of data records from 175 libraries describing various types of resources, including persons, books, authors, subjects, etc. The records were made available by building a straightforward RDF wrapper on top of the integrated library system.

Concerning Data Aggregators, one of the biggest efforts of publishing RDF data from aggregated data is *data.europeana.eu* [1], with its data source being the European Union's digital library Europeana[4]. There is an ongoing effort of making Europeana metadata available as LOD; however, scaling issues are again not addressed sufficiently and RDF stores are used for read-only access after an initial dump import. Moreover, the persistence of the URIs is a constant challenge in that approach: Despite having a robust URI design, Europeana is an aggregator and its collections are constantly being re-harvested, which leads to frequent changes of URIs.

4   Conclusion and Ongoing Work

We have presented the architecture that performs the efficient large-scale translation of the OA research metadata to Linked Data at http://lod.openaire.eu. We will extend this setup to interlink OpenAIRE with related datasets. While the implementation of the efficient incremental interlinking workflow is still in progress, we have already identified candidate datasets to interlink with and are in the process of determining rules to match OpenAIRE entities to entities in other datasets, and we created an initial collection of scholarly communication use cases that our interlinked datasets will support.

Acknowledgments. This work has been partially funded by the EU project OpenAIRE2020 (643410) and the DFG grant AU 340/9-1.

---

[4] http://www.europeana.eu.

References

1. Haslhofer, B., Isaac, A.: data.europeana.eu: The Europeana linked open data pilot. In: International Conference on Dublin Core and Metadata Applications, DCMI, pp. 94–104 (2011)
2. Haslhofer, B., Schandl, B.: The OAI2LOD server: exposing OAI-PMH metadata as linked data. In: Proceedings of the WWW 2009 Workshop on Linked Data on the Web, CEUR-WS, vol. 538 (2008)
3. Heath, T., Bizer, C.: Linked data: evolving the web into a global data space. Synth. Lect. Semant. Web Theory Technol. **1**(1), 1–136 (2011)
4. Papadakis, G., Alexiou, G., Papastefanatos, G., Koutrika, G.: Schema-agnostic vs schema-based configurations for blocking methods on homogeneous data. Proc. VLDB Endowment **9**(4), 312–323 (2015)
5. Summers, E., Isaac, A., Redding, C., Krech, D.: LCSH, SKOS and linked data. In: International Conference on Dublin Core and Metadata Applications, DCMI, pp. 25–33 (2008)
6. Vahdati, S., Karim, F., Huang, J.-Y., Lange, C.: Mapping large scale research metadata to linked data: a performance comparison of HBase, CSV and XML. In: Garoufallou, E., Hartley, R.J., Gaitanou, P. (eds.) MTSR 2015. CCIS, vol. 544, pp. 261–273. Springer, Cham (2015). doi:10.1007/978-3-319-24129-6_23
7. Vahdati, S., Lange, C., Alexiou, G., Papastefanatos, G.: Deliverable 8.2 - OpenAIRE LOD Services (2015)

A Framework for Keyphrase Extraction from Scientific Journals

Vidas Daudaravicius$^{(\boxtimes)}$

VTeX, Vilnius, Lithuania
vidas.daudaravicius@vtex.lt

Abstract. We present a framework for keyphrase extraction from scientific journals in diverse research fields. While journal articles are often provided with manually assigned keywords, it is not clear how to automatically extract keywords and measure their significance for a set of journal articles. We compare extracted keyphrases from journals in the fields of astrophysics, mathematics, physics, and computer science. We show that the presented statistics-based framework is able to demonstrate differences among journals, and that the extracted keyphrases can be used to represent journal or conference research topics, dynamics, and specificity.

Keywords: Keyphrase extraction · Journal · Collocation · TF-IDF

1 Introduction

Keyphrase extraction from single documents has been extensively examined for many years. Automatic keyphrase extraction concerns "the automatic selection of important and topical phrases from the body of a document" [20]. *Keyphrase extraction* is the selection of a set of phrases that are related to the main topics discussed in a given document. Document keyphrases have shown their potential for improving many natural language processing and information retrieval tasks. [9] presents a thorough survey of the state of the art in automatic keyphrase extraction, examining the major sources of errors made by existing systems and discussing the challenges ahead.

Recently, several shared tasks have been organized to evaluate the performance of various keyphrase extraction tools, including the ACL 2015 Workshop on "Novel Computational Approaches to Keyphrase Extraction" [8] and SemEval-2010 Task 5: "Automatic Keyphrase Extraction from Scientific Articles" [11]. The small number of large, publicly available data-sets of scientific texts with annotated keyphrases is a major difficulty in the keyphrase extraction research domain. Most of the data-sets for the keyphrase extraction task are not large enough [9]. For instance, [13] shows that adding a little additional training data improved the final results of the task [11], i.e., +7.4% for the F-score, raising it from 25.6 to 27.5. For the scientific domain, the data-sets amount to only a few hundred documents. This number is not sufficient to apply the widely used

© Springer International Publishing AG 2016
A. González-Beltrán et al. (Eds.): SAVE-SD 2016, LNCS 9792, pp. 51–66, 2016.
DOI: 10.1007/978-3-319-53637-8_7

TF-IDF measure, and it is difficult to avoid training overfitting. The arXiv.org database of scientific article preprints could be a very useful source of scientific articles to compile a data-set for the keyphrase extraction task. However, not many of the articles in the database have assigned keyphrases.

Table 1. Journals of the data-set.

Journal	Abbreviation	Documents	Tokens
Journal of Functional Analysis	JFAN	2490	26M
Journal of Algebra	JABR	4713	48M
Advances in Mathematics	AIMA	2041	28M
Solar Physics	SOLA	1683	11M
Journal of Optimization Theory and Applications	JOTA	1169	7M
Astrophysics and Space Science	ASTR	2880	12M
Annals of Operations Research	ANOR	1256	10M
Acta Applicandae Mathematicae	ACAP	799	5M
Total		17031	148M

Table 2. Data distribution by year.

Year	Documents	Tokens
2005	330	3M
2006	701	6M
2007	1603	14M
2008	1593	13M
2009	1608	14M
2010	1747	15M
2011	1950	17M
2012	1913	16M
2013	1564	12M
2014	2335	19M
2015	1687	16M
Total	17031	148M

Keyphrase extractions makes widespread use of multiword extraction techniques. [10] presents recent advances in multiword extraction. There are two main approaches for multiword extraction: syntax-based [10,15], and statistics-based [6,10]. [6] uses collocation segmentation to extract keyphrases from the Association for Computational Linguistics Anthology Reference Corpus (ACL ARC) [2] to study ACL history and research dynamics over the past 50 years. The distribution of keyphrases in the ACL ARC can be used to understand the main breakpoints of research across many years.

Little attention has been given to the extraction of keyphrases from larger sets of journals or conference papers with the aim to study research dynamics. Major conference organizers will often publish manually selected keyphrases from all accepted papers in proceedings prefaces, as a way to show trends in research. The goal of our study is to present a framework for keyphrase extraction from scientific journals in diverse research domains. While journal articles are often provided with manually assigned keywords, it is not clear how to extract statistically or syntactically significant keywords and measure their importance to the entire journal. In our study, we show that our statistics-based framework is able to demonstrate differences among journals, and can be used to represent journal or conference research, topics, dynamics, and specificity.

2   The Data-Set

The data-set used in our study has access to proprietary data from the VTeX production archive which is not publicly available. VTeX provides pre-publishing (copy-editing and typesetting) services to major science publishers for many years. All papers are LaTeX coded, even if some of them were originally submitted to a journal with other coding (e.g., MS Word). There were two initial requirements for the selection of journals: at least nine years of continuous typesetting at VTeX, and domain variety. We selected eight journals (see Table 1). Papers published between 2005 and 2015 in the fields of astrophysics, mathematics, physics, and computer science. Table 2 shows the yearly data distribution, which is evenly distributed except for the first two years. The journals publish different numbers of papers each year. The largest one is JABR, and the smallest one is ACAP. The total number of tokens in the corpus is 148 million. Although this data-set is far from the amount of data in reality, the size of the corpus is suficient to apply statistics and show results.

3   Framework Pipeline

The pipeline of the proposed framework comprises four main steps:

(1) text extraction (Sect. 3.1) and language detection (Sect. 3.2),
(2) candidate keyphrase list processing (Sect. 3.3),
(3) single article keyphrase weighting (Sect. 3.4), and, finally,
(4) smoothing of keyphrase weights for the sets of articles (Sects. 3.5 and 3.6).

3.1   LaTeX-to-Text conversion

We use the open-source tool `tex2txt`[1] for the conversion from LaTeX to text, because our source files are LaTeX-based. The tool is stand-alone and does not require any other LaTeX processing tools or packages. The primary goal of the tool is to extract the correct textual information from LaTeX files.

We need to point out that this tool makes some important changes to the original text: formal notation (i.e., mathematical expressions and other formulae) is substituted with the general category tag _MATH_. This substitution reduces the amount of interruptions of irrelevant language[2], and keeps the language of the article more coherent.

---

[1] See demo on-line: http://textmining.lt:8080/tex2txt.htm.
[2] In our case, this is mathematical language. Other cases may include a mix of English and French paragraphs in the same article.

3.2 Language Detection

Some papers in the selected journals were written in French with an abstract in English. We used a simple word-matching technique to detect articles written in English. We use two short word lists:

FR: de, et, le, une, sur, la, les, dans, est, pour;
EN: and, the, or, is.

An article is considered to be written in English if a text contains all of the words from the EN list, and none of words from the FR list. The technique is not universal, and useful only if both English and French languages are used. There are other statistical approaches to text language identification (see [1]).

3.3 Collocation Chains

Collocation segmentation is introduced in [7]. Collocation segmentation is a type of segmentation whose goal is to detect *collocated word sequences* and to segment a text into word sequences that we call *collocation chains*. Collocation chains can have any non-predefined length (even a single word). This definition differs from other collocation definitions that commonly use n-gram list-based approaches [3,16,19]. Collocation segmentation is related to collocation extraction using syntactic rules [12]. Syntax-based approaches allow us to extract collocations that are easier to describe, and the process of collocation extraction is well controlled. In our work, we use language-independent collocation segmentation for the data-set preprocessing, and the keyphrase candidate list is generated in a similar way as in [6].

Word Associativity. We use a Dice score to measure the associativity between two consecutive text tokens. Dice is defined as follows:

$$\text{Dice}(x_{i-1}; x_i) = \frac{2 \cdot \text{TF}(x_{i-1}; x_i)}{\text{TF}(x_{i-1}) + \text{TF}(x_i)},$$

where $\text{TF}(x_{i-1}; x_i)$ is the number of co-occurrences of x_{i-1} and x_i, and $\text{TF}(x_{i-1})$ and $\text{TF}(x_i)$ are the numbers of occurrences of x_{i-1} and x_i in the training corpus. If x_{i-1} and x_i tend to occur in conjunction, their Dice score will be high. The Dice score is sensitive to low-frequency word pairs (see the comparison of various associativity measures in [4]). If two consecutive words are used only once and appear together, there is a high chance that these two words are closely related and form some new concept, e.g., a proper name or a semantically closed term. A sequence of tokens is turned into a curve of Dice values between two adjacent tokens. This curve of associativity values is used to detect the boundaries of collocation chains.

Second-Order Derivative for Collocation Chain Boundaries. [5] introduces the average minimum law (AML) for setting collocation chain boundaries. Actually, AML is a second- order derivative, which is applied to three adjacent associativity values, and it is defined as follows:

$$
boundary(x_{i-1}, x_i) = \begin{cases} True \mid \text{Dice}(x_{i-2}; x_{i-1}) + \text{Dice}(x_i; x_{i+1}) - \\ \qquad\qquad -2 \cdot \text{Dice}(x_{i-1}; x_i) > 0 \\ False \mid \text{Otherwise.} \end{cases}
$$

If the second-order derivative value is positive, then two consecutive tokens are not joined into a collocation chain, and while two consecutive tokens are concatenated if the derivative value is negative.

Preprocessed Collocation Chains. The data-set contains 129,257 unique unigrams and 3,770,944 unique bigrams. We processed the data-set with collocation segmentation, and found 1,364,638 unique collocation chains. The list size of bigrams is twice the size that of collocation chains. List size grows quickly with the length of n-grams. Although collocation chains reduce the number of unique items, nevertheless n-gram features are preserved and many noisy bigrams and trigrams that occur only once are often omitted. The maximal length of collocation chains is 6 tokens and, which is similar as into [6], where the maximal chain length was 7 tokens. The average collocation chain length in the dictionary is 1.68 tokens. Collocation chains with 1, 2, or 3 tokens cover 99.5% of the corpus. In Table 3 we show collocation chains ending with the token *energy*. Chains starting or ending with verbs, determiners, numbers, or prepositions were removed from this list. The list exposes many different energy types and conceptions of energy, which givinge us a good sense of the variety and use of the term.

3.4 Keyphrase Weighting with TF-IDF and NTF-PIDF

Since 1972, when the inverse document frequency measure was introduced [17], the TF-IDF weighting method has been very successfully used in information retrieval and other natural language processing tasks. We use TF-IDF, which is defined as follows:

$$
\text{TF-IDF}(x) = \text{TF}(x) \cdot \ln\left(\frac{N}{\text{D}(x)}\right),
$$

where $\text{TF}(x)$ is the raw frequency of a term x in the data-set, N is the total number of articles in the data-set, and $\text{D}(x)$ is the number of articles where the term x occurs.

We also use the normalized probabilistic TF-IDF variation. We make normalize term frequency normalization against the article length and the average article length. The length of articles in the data-set varies from 246 to 88642 tokens. The weight of a term TF-IDF in a shorter article is much lower than the its weight of ain a longer article, even if the two articles are about the same topic.

Table 3. The list of collocation chains that end with *energy*.

accumulated ...	essentially different ...	laser ...	radiation ...
acoustic ...	excess ...	local ...	radiative ...
activation ...	excitation ...	low ...	relativistic ...
additional ...	exotic ...	lower ...	relativistic fermi ...
adiabatic ...	explosion ...	lowest ...	released ...
adm ...	extracted ...	magnetic ...	repulsive ...
alpha particle ...	false vacuum ...	mass and ...	rest ...
average ...	fermi ...	mass ...	rotational ...
beam ...	field ...	mass or ...	screening ...
binding ...	final ...	matter and ...	second ...
break ...	flare ...	matter ...	shear ...
burst ...	flowing ...	matter or ...	shock ...
calculation of ...	fluid ...	maximum ...	significant ...
cm ...	fraction of ...	mean ...	soliton ...
comparable ...	free ...	mechanical ...	solution and ...
compton ...	gas ...	minimum ...	source of ...
constant ...	graph of ...	missing ...	specific ...
correlation ...	gravitational binding ...	moller ...	spectral ...
cosmological nuclear ...	gravitational correla-	negative ...	standard ...
coulomb ...	tion ...	newtonian potential ...	state ...
cutoff ...	gravitational ...	nuclear ...	stored ...
cyclotron ...	gravitational wave ...	null ...	stress ...
dark ...	gravitomagnetic ...	orbital ...	strong ...
decaying vacuum ...	helmholtz ...	outburst ...	sufficiently high ...
dimensionless ...	high ...	particle ...	symmetry ...
dissipated ...	highest ...	peak ...	tev ...
dominant ...	holographic dark ...	phantom ...	thermal ...
dust matter ...	hydrodynamic explo-	phantom field ...	trace of ...
edge of ...	sion ...	photon ...	turbulence ...
effective ...	increase of ...	plasma ...	turbulent ...
effective plasma ...	increasing ...	plasmon ...	unit of ...
effective potential ...	infinite ...	position and ...	universe ...
elastic ...	initial ...	positive ...	vacuum ...
electric ...	instantaneous orbital ...	positron fermi ...	very high ...
electric potential ...	interaction ...	possible relativistic ...	vibration ...
electromagnetic ...	interaction potential ...	potential ...	wave ...
electron ...	internal ...	propagating wave ...	weak ...
electron thermal ...	intersystem correlation	pseudo...	wind ...
electrons with ...	...	quantum vacuum ...	zero ...
electrostatic ...	isotropic ...	radiant ...	
...	kinetic ...	radiated ...	
equipartition law of ...		radiates ...	

The term-occurrence counts in articles are proportional to the article length, therefore, we normalize frequencies to make them comparable. Such normalization is important when we compare keyphrases in separate articles, but though not within each article itselfindividually. The term-frequency normalization is as follows:

$$\text{NTF}(x) = \text{TF}(x) \cdot \frac{\text{avgDocLen}}{\text{length}(D_x)},$$

where avgDocLen is the average article length in the data-set, and $\text{length}(D_x)$ is the length of an article with term x. The average article length in the data-set is 9090 unigram tokens and 5655 collocation chain tokens. In general, the average article length is constant and is not necessary in calculation, i.e., the constant becomes equal to 1.

IDF probabilistic variation is discussed in [14]. The probabilistic variation of IDF we use is defined as follows:

$$\text{PIDF}(x) = \ln\left(\frac{N - \text{D}(x) + 1}{\text{D}(x) + 1}\right).$$

In the case of a term which occurs in more than half of the articles in the data-set, the formula defines a negative weight. This is a somewhat odd prediction for a term. In practice, all terms with negative weight are stop-words or function words[3]. The normalized probabilistic TF-IDF is defined as follows:

$$\text{NTF-PIDF} = \text{NTF} \cdot \text{PIDF}.$$

TF-IDF and NTF-PIDF are applied to each term of each article in the data-set. We take the top 100 most significant terms of each article for the next steps described in the following sections.

3.5 Journal Keyphrases

We have described keyphrase extraction from articles in the sections above. In this section we extend keyphrase extraction from a single article to subgroups of larger sets of articles. Articles can be grouped by year and/or by journal, and/or by other categories. One straightforward way to extract keyphrases from groups of large data-sets is to calculate the term weights for each group separately. While the term frequency of a group is easy to calculate, it is not clear what should be the number of articles for each term occurrence. Typically, a journal issue contains from 10 to 50 articles. Such a small subgroup is not sufficient for TF-IDF weighting, and we will not be able to extract keyphrases properly for each separate journal, or yearly journal article groups. To tackle the problem of document count, we calculate the average of TF-IDF of article terms in each subgroup as follows:

$$\text{TF-IDF}_{\text{AVG}}(x|g) = \frac{\sum \text{TF-IDF}(x, g)}{\text{D}(x, g)},$$

where $\sum \text{TF-IDF}(x, g)$ is the sum of TF-IDF(x) terms of articles in a subgroup g, and $\text{D}(x, g)$ is the number of articles in the subgroup g where the term x occurs. The average of NTF-PIDF is calculated similarly:

$$\text{NTF-PIDF}_{\text{AVG}}(x|g) = \frac{\sum \text{NTF-PIDF}(x, g)}{\text{D}(x, g)}.$$

---

[3] Function words are words that have little lexical meaning or have ambiguous meaning, but instead serve to express grammatical relationships with other words within a sentence (https://en.wikipedia.org/wiki/Function_word). For instance, *and, or, the,* and *a* are all function words.

Table 4. Top keyphrase lists of ASTR. Grey highlights keyphrases that occur in both top lists of two different measures.

| Sigmoid additive smoothing | | No smoothing | |
NTF-PIDF$_{ADD}$	TF-IDF$_{ADD}$	TF-IDF$_{AVG}$	NTF-PIDF$_{AVG}$
periodic orbits	periodic orbits	gsxr flares	gsxr flares
black hole	equilibrium points	cemp	proxima
equilibrium points	black hole	outgrowths	bal quasars
brane	sn ia	bexb	emp
bulk viscosity	brane	fz ori	pbps
chaplygin gas	primaries	spin spacecraft	center manifold
bulk viscous	triangular points	pbps	cemp
dusty plasma	jet	proxima	smf
dark energy	cluster	finger	comet holmes
primaries	body problem	nutation damper	flyer
jet	co	lp uma	tv columbae
triangular points	disk	gclf	string ball
positrons	bulk viscosity	ba stars	spin spacecraft
shock waves	dark energy	magnetosonic critical curve	dsphs
scalar field	dusty plasma	issv solutions	ba stars
co	scalar field	stereo pairs	strong spes
cluster	oblateness	shaped fingers	nutation damper
solitary waves	planet	bal quasars	u geminorum
bianchi type	shock waves	center manifold	information bits
body problem	disc	bf	rgda
apparent horizon	star formation	cehe	wz sge
alma	bl lacs	pywifes	soft excess
oblateness	globular clusters	rgda	triple asteroids
gamma	asteroid	cyclical universe	ao psc
de sitter	positrons	smf	diamond detector
entropy	chaplygin gas	fingers	fec
growth rate	de sitter	periodic modes	electron hole
universe	growth rate	fyris alpha	submm ebl
neutron star	equilibrium point	attitude stability	gm cep
event horizon	solitary waves	expansive homogeneous	bf
star formation	moon	matter objects	qq vul
restricted three	neutron star	u geminorum	lmt
black holes	bulk viscous	isotropic relativistic universe	rv psc
holographic dark energy	shell	preferred alignment	primary cmes
mach number	positron	outgrowth	ulp cepheids
negative ions	light curves	protons flares	w ser
well behaved	mass loss	dibs	narrow cmes
positron	clusters	information bits	umfs
ion acoustic	galaxies	barium stars	ux mon
double layers	holographic dark energy	sxr emissions	bellert
hot electrons	magnetic field	edod	sterile neutrinos
disk	eos parameter	koi	capella
clusters	restricted three	dsphs	gclf
wso	apparent horizon	lunar wake	nqda
hawking radiation	double layers	w ser	giant pulses
dust grains	blazars	smgs	aloh
radiation pressure	universe	issv	nel rates
thermodynamics	mach number	neutral formaldehyde	tev j
blazars	metallicity	ux mon	ircss
disc	h	polarization force	bok globule

An example of the top 50 extracted keyphrases from ASTR is shown in Columns 3 and 4 of Table 4. The most significant keyphrase is *gsxr flares*, using both weighting measures. GSXR is an acronym of the term *great soft x-ray*. The extracted keyphrase lists with the two weighting measures do not significantly correlate. However, the keyphrase *gsxr flares* is used frequently in only one article.

An expanded *gsxr flares* keyphrase is used only once, and it occurs in the title of that same article. What are the true keyphrases for the ASTR journal? Can we accept the *gsxr flares* keyphrase, which is used frequently in only one article, as a descriptor of the entire ASTR journal for the 2005–2015 period? In the following section we present a solution to this problem using *additive smoothing*.

We also noticed that keyphrases extracted with TF-IDF$_{\text{AVG}}$ weighting are much longer than the keyphrases extracted with NTF-PIDF$_{\text{AVG}}$. This is due to a high correlation between term frequency and term length (see [18]). NTF-PIDF$_{\text{AVG}}$ uses term frequency normalization to reduce the impact of term frequency to the term significance value. Short and frequent terms are more abstract than longer and less frequent ones. Therefore, this property can be used to extract either more abstract or more detailed terms, which might depend on the task.

3.6   Additive Smoothing of TF-IDF

Smoothing is widely used to reduce noise and irregularities in data series, or to smooth categorical data. Additive smoothing is a common approach in statistical language modeling. The aim of applying smoothing is to reduce probabilistic irregularities. In Sect. 3.5 we showed that the average of TF-IDF is not sufficient for the extraction of keyphrases for larger sets of articles. If a term is used in a few articles only but its significance is high, then its average significance will also be high. Our goal is to extract keyphrases that represent entire set of articles. Therefore, we need some balance between article term significance and the number of articles in which this term occurs. The larger the number of articles with a term x and the higher the significance of the term x in these articles, the more this term is significant for the whole subset of articles. We implement this approach by applying additive smoothing to the TF-IDF$_{\text{AVG}}$ calculation. This approach reduces the average of TF-IDF$_{\text{AVG}}$ regarding the number of articles in which a keyphrase is used. Thus, the fewer the number of articles, the greater the amount of additive smoothing. After some manual experimentation, we adjusted the following *sigmoid* function:

$$\text{sigmoid}(i) = \frac{200}{1 + e^{-(-0.05) \cdot i}},$$

where i is the number of articles in which a term x occurs.

This gives us a high smoothing value to terms that occur in a few articles only, and a lower smoothing value to terms that occur in many articles. For instance, if a term occurs in only one article then the smoothing parameter is equal to 97; for 50 or 100 articles, respectively, the parameter drops to 15 and 1.3. We use the following additive smoothing formulas:

$$\text{TF-IDF}_{\text{ADD}}(x) = \frac{\text{sigmoid}(\text{D}(x)) + \sum \text{TF-IDF}(x)}{\text{sigmoid}(\text{D}(x)) + \text{D}(x)},$$

$$\text{NTF-PIDF}_{\text{ADD}}(x) = \frac{\text{sigmoid}(D(x)) + \sum \text{NTF-PIDF}(x)}{\text{sigmoid}(D(x)) + D(x)}.$$

The results of additive smoothing (see Table 4) show high correlation for the top 50 extracted keyphrases using both weighting measures. For instance, the keyphrase *periodic orbits* occurs in 94 articles of ASTR, and its frequency is 1397; *black hole*: 562 and 6184; *dark energy*: 544 and 5073; and *equilibrium points*: 134 and 1684. We do not count instances when a keyphrase is nested in another term; for example, if nesting is considered, then the *black hole* term counts are 637 and 8063. In Table 4 we see *co* as a keyphrase, which is ambiguous in ASTR, as it can stand for either *cobalt* or *carbon monoxide*. Both meanings are extensively used in articles. Two-word keyphrases are the most common in journals, except SOLA and ANOR (see Table 5). The top keyphrase list of SOLA is full of abbreviations and acronyms. The top list of ANOR contains many single-word keyphrases. Keyphrase lists extracted with (TF-IDF$_{\text{ADD}}$ and NTF-PIDF$_{\text{ADD}}$) correlate. The question of how this correlation can help get even higher keyphrase extraction accuracy could be answered in a following study.

At this point, we have extracted keyphrases from a set of articles. In the following sections, these keyphrases will be used to analyze the main research trends in particular sets of articles.

4   Journal Keyphrases

In this and the following sections we discuss possible use cases for extracted keyphrases. In this section we analyze the differences among journals in our data-set.

We can find the publishing topics of journals on the Internet. Often these descriptions overlap considerably. We found the following research subjects of journals:

- *Journal of Functional Analysis* (JFAN) related subjects[4]: Significant applications of functional analysis, including those to other areas of mathematics; New developments in functional analysis; Contributions to important problems in and challenges to functional analysis.
- *Journal of Algebra* (JABR) related subjects[5]: Results obtained by computer calculations; Classifications of specific algebraic structures; Description and outcome of experiments; Papers emphasizing the constructive aspects of algebra, such as description and analysis of new algorithms; Interactions between algebra and computer science, such as automatic structures, word problems, and other decision problems in groups and semigroups; Contributions are welcome from all areas of algebra, including algebraic geometry or algebraic number theory.

---

[4] http://www.journals.elsevier.com/journal-of-functional-analysis/.
[5] http://www.journals.elsevier.com/journal-of-algebra/.

Table 5. The top 50 list of extracted keyphrases of different journals. *Keyphrases with strikeouts cannot really be used as keyphrases. We show all automatically extracted keyphrases in order of significance. We omit significance values here and later to save space.*

AIMA	convex body, dg, convex bodies, hopf algebra, operad, polytope, weak equivalence, cell, cells, stack, groupoid, intersection body, ample, hausdorff dimension, tree, weak equivalences, spectral sequence, functor, cocycle, homotopy, moduli space, intersection bodies, quiver, line bundle, category, geodesic, scalar curvature, vector bundle, symmetric monoidal, sheaf, simplex, initial data, block, von neumann, edges, graph, triangulated category, monad, orbifold, simplices, bundle, volume, graded, hypersurface, gorenstein, simplicial, finite type, morphism, simplicial complex, semistable
JABR	vertex operator, lie superalgebra, vertex algebra, superalgebra, hopf algebra, crystal, defect group, fusion system, grobner basis, koszul, jordan algebra, definable, numerical semigroup, permutable, hilbert function, locally nilpotent, braided, representation type, bialgebra, lie superalgebras, soluble, engel, subnormal, leavitt path, almost split, del pezzo, tableau, monomial ideal, hopf algebras, dimension vector, central simple, gorenstein, betti numbers, cofinite, complete intersection, inner ideal, ample, quiver, clean, comodule, pi, local cohomology, polycyclic, artin algebra, semiprime, lie algebra, supersolvable, block, line bundle, supersoluble
ACAP	random variables, differential equations, initial data, boundary conditions, vector field, weak solution, vector fields, periodic solutions, system, hamiltonian, boundary value, positive solution, positive solutions, differential equation, h, nonlinear, operator, estimator, solutions, symmetries, stokes equations, pdes, ~~solution of~~, graph, distribution, functional, initial conditions, hopf bifurcation, conservation laws, species, fixed point, periodic solution, hpm, curve, estimators, fluid, asymptotically stable, global existence, symmetry, lie algebra, positive constant, differential invariants, equilibrium point, ham, prolongation, reaction, operators, banach space, zeros, population
ANOR	job, supply chain, jobs, machine, queue, game, retailer, customers, customer, dmu, machines, supplier, coalition, players, local search, player, server, policy, operations research, schedule, dea, portfolio, processing times, manufacturer, items, servers, risk, inventory level, column generation, criteria, markov chain, facility, agent, makespan, inventory, processing time, service time, master problem, benders decomposition, stage, node, patients, dmus, period, resource, special volume, completion time, buffer, graph, network
ASTR	periodic orbits, black hole, equilibrium points, brane, bulk viscosity, chaplygin gas, bulk viscous, dusty plasma, dark energy, primaries, jet, triangular points, positrons, shock waves, scalar field, co, cluster, solitary waves, bianchi type, body problem, apparent horizon, alma, oblateness, gamma, de sitter, entropy, growth rate, universe, neutron star, event horizon, star formation, restricted three, black holes, holographic dark energy, mach number, negative ions, well behaved, positron, ion acoustic, double layers, hot electrons, disk, clusters, wso, hawking radiation, dust grains, radiation pressure, thermodynamics, blazars, disc
JOTA	~~ref~~, optimal control, game, nash equilibrium, player, value function, maximal monotone, quasiconvex, valued mapping, constraint qualification, delay, primal, variational inequality, pseudomonotone, p, lsc, optimality conditions, strongly monotone, efficient solution, optimal solution, central path, algorithm, stationary point, global minimizer, variational inequalities, players, stochastic, multifunction, firm, augmented lagrangian, inequality constraints, lower semicontinuous, cost function, control, usc, upper semicontinuous, locally lipschitz, subdifferential, differential game, lmi, euclidean jordan, convex subset, duality gap, convex, normal cone, lmis, banach space, necessary optimality conditions, newton method, proposed method
SOLA	hi, filament, mc, flux rope, mcs, ca ii, solar wind, cme, type ii, icmes, icme, sep events, meridional flow, type iii, ar, prominence, flux tube, rotation rate, sunspot groups, gcr intensity, sunspot number, sunspot, type iii bursts, filaments, cmes, burst, hard x, coronal holes, he ii, hcs, hmi, ips, flare, radio bursts, fe i, coronal hole, loop, sunspot numbers, time series, eis, flux emergence, shock, cor, ars, solar activity, eruption, current sheet, tsi, solar cycle, sunspots
JFAN	toeplitz operators, completely positive, von neumann, dirichlet form, heat kernel, composition operator, composition operators, ground state, semigroup, completely bounded, poincare inequality, banach space, brownian motion, banach algebra, h, amenable, ricci curvature, invariant subspace, approximation property, almost surely, weak solution, initial data, locally convex, nuclear, critical point, posedness, unital, fixed point, lipschitz domain, cocycle, inner function, carleson measure, locally compact, metric space, toeplitz operator, weakly compact, graph, sobolev inequality, hilbert, representing measure, positive solution, invariant subspaces, positive definite, crossed product, operator space, fredholm, banach, strongly continuous, stokes equations, ergodic

- *Advances in Mathematics* (AIMA) related subjects[6]: Emphasizing contributions that represent significant advances in all areas of pure mathematics.
- *Acta Applicandae Mathematicae* (ACAP) related subjects[7]: Classical Continuum Physics, Complexity, Computer Science, Mathematics, Theoretical, Mathematical, and Computational Physics.

The differences among the listings of related subjects are fuzzy. The topics covered by JFAN, JABR, AIMA, and ACAP overlap. Whether or not a submitted article is accepted depends on how successfully it conforms to the journal research subjects. There is a high chance that a submitted article will be rejected if the article is not in line with the journal research subjects. Journal keyphrases from at least the past five years would definitely help authors to choose the most appropriate journal for submission. Table 5 shows the top keyphrases for each journal in the data-set. The lists of keyphrases show that, in fact, all four journals focus on different topics, and there is no intersection among the main terms of each journal, making it easier to get a sense of the specificities of these journals.

5   Research Dynamics

In this section, we analyze the *Astrophysics and Space Science* (ASTR) journal. ASTR is an international journal of astronomy, astrophysics, and space science. ASTR related subjects are as follows: Astrobiology; Astronomy, Observations, and Techniques; Astrophysics and Astroparticles; Cosmology; Extraterrestrial Physics, Space Sciences. The description of the journal is as follows[8]:

> "*Astrophysics and Space Science* publishes original contributions and invited reviews covering the entire range of astronomy, astrophysics, astrophysical cosmology, planetary and space science and the astrophysical aspects of astrobiology. [...] We particularly welcome papers in the general fields of high-energy astrophysics, astrophysical and astrochemical studies of the interstellar medium including star formation, planetary astrophysics, the formation and evolution of galaxies and the evolution of large scale structure in the Universe. [...]"

In Table 4 in Column 1, we show the top 50 most significant keyphrases of ASTR journal. The keyphrases describe the main terms of the journal for the past 10 years. As we expected, the terms *periodic orbits*, *black hole*, *equilibrium points*, and *dark energy* are on the top 10 list. The top 10 list shows the recent research trends of ASTR.

The top keyphrases of ASTR for each year from 2007 to 2015 are shown in Table 6. The term *star formation* is significant up to 2011, while the term

---

[6] http://www.journals.elsevier.com/advances-in-mathematics/.

[7] http://www.springer.com/mathematics/journal/10440.

[8] http://www.springer.com/astronomy/astrophysics+and+astroparticles/journal/10509.

Table 6. The top 50 extracted keyphrases from ASTR each year, 2007–2015.

2007	jet, gamma, neutron star, magnetic field, pulsar, ray emission, rx j, neutron stars, rays, crust, black hole, star, field equations, sources, pulsars, source, tev, glast, ray sources, psr b, disk, blazars, ray, energy density, high energy, axps, radio pulsars, emission, universe, hard x, radio emission, radio, kev, accretion rate, magnetic fields, g, flux, vhe, polar cap, accretion disk, energy, psr j, cosmic rays, egret, wind, grb, agn, angular momentum, light curve, scalar field
2008	alma, universe, co, field equations, star formation, bianchi type, mass, energy density, molecular gas, einstein, galaxies, cosmological models, scalar field, hcn, perfect fluid, h, magnetic field, gas, equilibrium points, star, metric, code, cosmological model, general relativity, dust, sma, molecules, ~~km s~~, stars, molecular clouds, disks, angular momentum, primaries, chemistry, emission, galaxy, stellar models, black hole, periodic orbits, gravitation, radiation pressure, cn, dark energy, ch, aca, disk, deceleration parameter, body problem, light curve, cm
2009	cluster, clusters, stars, universe, star clusters, star formation, galaxies, field equations, energy density, magnetic field, galaxy, o vi, dark energy, mass, uv, massive stars, ~~km s~~, cosmological constant, gas, star, mass segregation, cm, black hole, age, plasma, m, hii regions, h, metallicity, ly, cluster mass, experiment, ages, stellar, jet, mag, chaplygin gas, bianchi type, experiments, ns, shock, orbital period, disk, virial equilibrium, hst, flyer, antennae, ism, perfect fluid, uvot
2010	black hole, universe, stars, star, hot subdwarfs, sdb stars, dark energy, magnetic field, field equations, energy density, scalar field, cosmological constant, hot subdwarf, helium, mass, quasinormal frequencies, metallicity, solar, orbital period, black holes, solitary waves, einstein, modes, white dwarf, mass transfer, age, sun, main sequence, sdbs, general relativity, convective core, galaxies, binaries, galaxy, scale factor, kinetic energy, sdo stars, convection, mass loss, dh, cmes, sdb, angular momentum, accelerated expansion, binary, globular clusters, bulk viscosity, cosmological models, event horizon, white dwarfs, models
2011	universe, black hole, dark energy, magnetic field, energy density, field equations, neutron star, star, event horizon, chaplygin gas, solitary waves, scalar field, periodic orbits, stars, cosmological constant, pressure, perfect fluid, galaxies, plasma, mass, wso, apparent horizon, maximum mass, dust, well behaved, thermodynamics, deceleration parameter, positrons, dark matter, scale factor, general relativity, temperature, red shift, star formation, sound, dust grains, uv, galaxy, galex, gsxr flares, particle, ~~equation of~~, quintessence, electron, metric, gas, energy, hubble parameter, angular momentum, positron
2012	black hole, dark energy, universe, magnetic field, solitary waves, scalar field, gravity, entropy, field equations, dusty plasma, energy density, cosmological constant, mass, ions, eos parameter, dark matter, plasma, dust, electron, soliton, positrons, dust grains, ion, general relativity, restricted three, stars, black holes, brane, body problem, cepheids, ~~scale factor~~, horizon, periodic orbits, positron, electrons, deceleration parameter, dls, shock waves, star, hot electrons, ion acoustic, number density, perfect fluid, amplitude, bianchi type, electric field, thermodynamics, event horizon, galaxies, ~~well behaved~~
2013	black hole, dark energy, universe, solitary waves, dusty plasma, magnetic field, black holes, entropy, energy density, scalar field, event horizon, field equations, dust, cosmological constant, horizon, gravity, ~~scale factor~~, periodic orbits, dispersion relation, deceleration parameter, ion acoustic, ions, electron, shock waves, dark matter, amplitude, positrons, mass, equilibrium points, plasma, ~~eqs~~, bulk viscous, positron, bulk viscosity, body problem, double layers, primaries, solitons, brane, solitary wave, de sitter, eos parameter, dust grains, hubble parameter, electrons, reductive perturbation, negative ions, growth rate, spectral index, da
2014	black hole, dark energy, equilibrium points, universe, body problem, primaries, scalar field, magnetic field, solitary waves, dusty plasma, dust, field equations, gravity, dark matter, oblateness, ~~restricted three~~, energy density, triangular points, ~~scale factor~~, motion, plasma, deceleration parameter, spectral index, holographic dark energy, ion acoustic, ions, hubble parameter, equilibrium point, radiation pressure, mass, chaplygin gas, electron, bianchi type, wave, black holes, infinitesimal mass, electrons, ~~eqs~~, positrons, growth rate, cosmological constant, ~~number density~~, dispersion relation, ~~equation of~~, eos parameter, redshift, ion, galaxies, solitary wave, dust grains
2015	black hole, dark energy, gravity, universe, scalar field, field equations, energy density, star, magnetic field, hubble parameter, eos parameter, de sitter, dark matter, bianchi type, cosmological constant, equilibrium points, gr, inflation, asteroid, ~~equation of~~, deceleration parameter, body problem, ~~eqs~~, black holes, radial pressure, primaries, stars, einstein, psr j, motion, modified gravity, momentum tensor, fluid, general relativity, compact stars, perfect fluid, event horizon, quintessence, model, metric, electric field, eos, spacetime, spherically symmetric, anisotropic, gravitational collapse, disk, accelerated expansion

magnetic field is used constantly over many years. *Black hole* becomes one of the most significant terms after 2010. In the Wikipedia article about black holes, we find the following description, which explains the rise of *black hole*[9]:

> "... black holes do not directly emit any signals other than the hypothetical Hawking radiation; [...] A possible exception to the Hawking radiation being weak is the last stage of the evaporation of light (primordial) black holes; searches for such flashes in the past have proven unsuccessful [...]. NASA's Fermi Gamma-ray Space Telescope launched in 2008 will continue the search for these flashes."

It is likely that this new gamma-ray telescope had a significant impact on research into black holes. Evidence for this conclusion can be found in the trending significance of the keyphrase *black hole* across many years of ASTR. In less than two years, researchers began collecting data from the new telescope and publishing their new discoveries.

This year-by-year comparison of the keyphrases of ASTR shows research trends, dynamics, and breakthroughs in astrophysics. Journals could publish such keyphrase lists with significance values as supplementary data either as an appendix in each volume, or yearly, significantly assisting potential contributors in assessing where best to submit their research for publication.

6   Discussion and Future Work

We have presented our framework for keyphrase extraction from sets of scientific articles. The framework differs from similar commercial tools, e.g., JANE[10] and HelioBLAST[11]. The HelioBLAST text similarity engine finds text records that are similar to the submitted query. JANE (Journal/Author Name Estimator) uses the short text of an article (e.g., the title and/or abstract) and searches for journals, authors, or articles. JANE compares the document to millions of documents in the MEDLINE database. Both tools are article-based approaches. The similarity distance between query and journal is based on the accumulated similarity between query and articles.

The properties of our framework are the following:

- The extracted keyphrases are journal-dependent and the direct connections to articles are dropped. It lifts up keyphrase lists to more general journal representation.
- The extracted keyphrases allow queries to be compared to journals instead of only articles. It allows query processing to be sped up considerably.
- The extracted keyphrases expose more general representations of journals than sets of articles. The framework is simple to implement and can be adopted and used independently for separate journals. It is flexible enough that centralized databases are not substantial.

---

[9] https://en.wikipedia.org/wiki/Black_hole.
[10] http://jane.biosemantics.org/.
[11] http://helioblast.heliotext.com/.

A list of keyphrases can be useful as supplementary material for the following (albeit incomplete) list of tasks:

- It can support describing the main research objectives of a journal and can help the research community to follow the main research trends and changes.
- It can support deciding whether a newly submitted article conforms with a journal's topics and can help researchers to choose the most appropriate journal to which to submit a new article.
- It can help to evaluate whether a topic has been extensively studied, is a new trend, or is the revival of an old topic.
- It can help libraries and search systems to index journals, and make queries more accurate.

The lack of manually annotated data is a formidable barrier for a thorough evaluation of the quality of extracted journal keyphrases. While we can evaluate the precision of extracted keyphrases (i.e., how accurately we selected them), we cannot easily evaluate recall (i.e., have we selected all of the important ones). In the near future, we plan to involve several editorial boards to evaluate the framework's quality and relevance.

The proposed framework only uses statistics, and no language-dependent tools are necessary to apply this framework. Therefore, the framework can be applied to new journals and new languages without specifically requiring language-dependent tools.

References

1. Baldwin, T., Lui, M.: Language identification: the long and the short of the matter. In: Human Language Technologies: The 2010 Annual Conference of the NAACL, Los Angeles, CA, pp. 229–237 (June 2010)
2. Bird, S., Dale, R., Dorr, B., Gibson, B., Joseph, M., Kan, M.Y., Lee, D., Powley, B., Radev, D., Tan, Y.F.: The ACL anthology reference corpus: a reference dataset for bibliographic research in computational linguistics. In: Proceedings of the Language Resources and Evaluation Conference (LREC 2008), Marrakesh, Morocco, May 2008
3. Choueka, Y.: Looking for needles in a haystack, or locating interesting collocational expressions in large textual databases. In: Proceedings of the RIAO Conference on User-Oriented Content-Based Text and Image Handling, pp. 21–24. Cambridge, MA (1988)
4. Daudaravicius, V., Marcinkeviciene, R.: Gravity counts for the boundaries of collocations. Int. J. Corpus Linguist. $9(2)$, 321–348 (2004)
5. Daudaravicius, V.: The influence of collocation segmentation and top 10 items to keyword assignment performance. In: Gelbukh, A. (ed.) CICLing 2010. LNCS, vol. 6008, pp. 648–660. Springer, Heidelberg (2010). doi:10.1007/978-3-642-12116-6_55
6. Daudaravicius, V.: Applying collocation segmentation to the ACL anthology reference corpus. In: Proceedings of the ACL-2012 Special Workshop on Rediscovering 50 Years of Discoveries, Jeju Island, Korea, pp. 66–75, July 2012
7. Daudaravicius, V.: Collocation segmentation for text chunking. Ph.D. thesis. Vytautas Magnus University, January 2013

8. Gollapalli, D.S., Caragea, C., Li, X., Giles, L.C.: Proceedings of the ACL 2015 Workshop on Novel Computational Approaches to Keyphrase Extraction (2015)
9. Hasan, K.S., Ng, V.: Automatic keyphrase extraction: a survey of the state of the art. In: Proceedings of the 52nd Annual Meeting of the Association for Computational Linguistics, pp. 1262–1273, Baltimore, Maryland, June 2014
10. Kilgarriff, A., Rychly, P., Kovar, V., Baisa, V.: Finding multiwords of more than two words. In: Proceedings of the 15th EURALEX International Congress, Oslo, pp. 693–700 (2012)
11. Kim, N.S., Medelyan, O., Kan, M.Y., Baldwin, T.: SemEval-2010 task 5: automatic keyphrase extraction from scientific articles. In: Proceedings of the 5th International Workshop on Semantic Evaluation, pp. 21–26 (2010)
12. Lin, D.: Extracting collocations from text corpora. In: First Workshop on Computational Terminology, Montreal (1998)
13. Lopez, P., Romary, L.: HUMB: automatic key term extraction from scientific articles in GROBID. In: Proceedings of the 5th International Workshop on Semantic Evaluation, Uppsala, Sweden, pp. 248–251, July 2010
14. Robertson, S.: Understanding inverse document frequency: on theoretical arguments for IDF. J. Documentation **60**, 503–520 (2004)
15. Seretan, V.: Syntax-Based Collocation Extraction. Text, Speech and Language Technology, vol. 44. Springer, Netherlands (2011)
16. Smadja, F.: Retrieving collocations from text: Xtract. Comput. Linguist. **19**, 143–177 (1993)
17. Spärck Jones, K.: A statistical interpretation of term specificity and its application in retrieval. J. Documentation **28**, 11–21 (1972)
18. Strauss, U., Grzybek, P., Altmann, G.: Word length and word frequency. In: Grzybek, P. (ed.) Contributions to the Science of Text and Language: Word Length Studies and Related Issues, vol. 31, pp. 277–294. Springer, Netherlands (2006)
19. Tjong Kim Sang, E.F., Buchholz, S.: Introduction to the CoNLL-2000 shared task: Chunking. In: Proceedings of CoNLL-2000 and LLL-2000, Lisbon, Portugal, pp. 127–132 (2000)
20. Turney, P.D.: Learning algorithms for keyphrase extraction. Inf. Retrieval **2**(4), 303–336 (2000)

Building Scholarly Data Forest

Marko Požega, Dario Poljak, and Kristina Kocijan[✉]

Department of Information and Communication Sciences,
Faculty of Humanities and Social Sciences,
University of Zagreb, Zagreb, Croatia
{mpozegal,dpoljakl,krkocijan}@ffzg.hr

Abstract. In this paper, we will demonstrate syntactic analysis and visualization of scientific data, namely references from scientific papers. Our main goal is to build a parser which could extract references from scientific papers, convert them to XML format, send to custom visualization algorithm and present in a web interface as a *ReferenceTree* for a single author. For this process, we use several different technologies such as NLP software NooJ, programming languages PHP and JavaScript in combination with HTML5. Our main problem was dissimilarity in reference styles between articles. Thus, our parser was designed to recognize different reference source (book, paper, web page) in APA, MLA and Chicago reference styles. As for the visualization idea, we have chosen the concept of presenting an author as a tree, the publication years as the main branches, the articles/books as twigs and references used in each article/book as the leaves. The books are grouped on the left side of the tree while the articles are grouped on the right side. With final output, every processed author should have a unique tree (preferences of references) and could be compared with the rest of the scientific forest.

Keywords: Scholarly data · Network visualization · Contact trees · Egocentric networks · NLP · APA · MLA · Chicago reference style · Science mapping · *ReferenceTree*

1   Introduction and Related Work

When we talk about the tree view of a network type data it is very often that we are talking about the trees that are using connected nodes with either a top-down or right-to-left orientation. Sometimes this so called 'tree view' has a rather circular shape network or matrix. Although informative breadth-wise, such visualizations are usually very modest in the depth of information they are able to show. It is quite recently that a new tree view visualization has been proposed [1] with a (real) tree shape visualization called ContactTrees (since they were originally designed to show person's social ties). Such trees have an egocentric approach with the ability to show multilevel aspects of social interactions in just a glance [1–5] that may be of help to sociologists as well as data managers as suggested by [2].

Our work is very much inspired by the work presented in [2]. We applied a similar approach in building our scientific reference trees which we present here as a new tool for science mapping as defined in [6]. However, our main concern is to show which

© Springer International Publishing AG 2016
A. González-Beltrán et al. (Eds.): SAVE-SD 2016, LNCS 9792, pp. 67–72, 2016.
DOI: 10.1007/978-3-319-53637-8_8

papers an author has cited throughout his/her academic career, rather than to visualize scientific disputes among different authors or their co-publishing behaviors.

In the sections that follow we will explain in more details steps involved in building *ReferenceTrees* starting with the data and an NLP tool we used for building syntactic grammars for automatic recognition and classification of references and finishing with the more detailed description of a tree. We will conclude the paper with some additional future work ideas.

2   Recognizing the References

We built syntactic grammar for reference recognition with an NLP tool - NooJ[1] constructed by Silberztein [7]. NooJ provides a graphical editor for building powerful syntactic grammars (graphs) that are well suited for our purpose. It allows us to create functional but also visually understandable grammars (Fig. 1). Each graph uses nodes that can be NooJ or regular expressions, plain text or even variables. It also uses the strength of a transducer and enables us to produce customized output such as XML like notation of the data needed for our *ReferenceTrees* (Fig. 3).

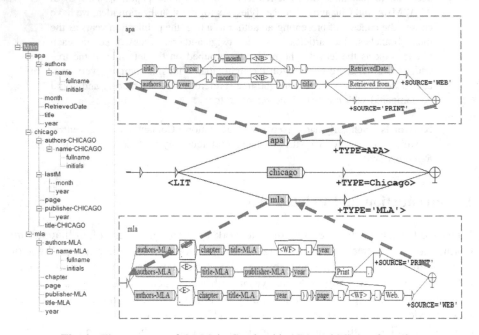

Fig. 1. The structure of the Main Graph with APA and MLA subgraphs

[1] NooJ can freely be downloaded from http://www.nooj4nlp.net/.

chicago

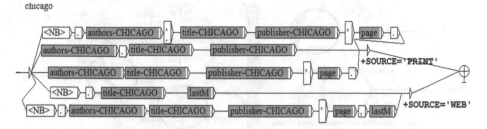

Fig. 2. Subgraph for recognition of Chicago style reference

Seq.

‹LIT+AUTHOR1='Agi \, Ž. '+AUTHOR2='Tadi \, M. '+YEAR='2006 '+TITLE='Evaluating Morphosyntactic Tagging of Croatian Texts '+SOURCE='PRINT'+TYPE='APA›
‹LIT+AUTHOR1='Agi \, Ž. '+AUTHOR2='Tadi \, M. '+AUTHOR3='Dovedan\, Z. '+YEAR='2008 '+TITLE='Combining part-of-speech tagger and inflectional lexicon for Croatian '+SOURCE='PRINT'+TYPE='APA›
‹LIT+AUTHOR1='Brants\, T. '+YEAR='2000 '+TITLE='TnT – A Statistical Part-ofSpeech Tagger '+SOURCE='PRINT'+TYPE='APA›
‹LIT+AUTHOR1='Erjavec\, T. '+YEAR='2004 '+TITLE='Multexf-East Version 3: Multilingual Morphosyntactic Specifications\, Lexicons and Corpora '+SOURCE='PRINT'+TYPE='APA›
‹LIT+AUTHOR1='Halácsy\, P. '+AUTHOR2='Kornai\, A. '+AUTHOR3='Oravecz\, C. '+YEAR='2007 '+TITLE='HunPos |- an open source trigram tagger '+SOURCE='PRINT'+TYPE='APA›
‹LIT+AUTHOR1='Rabiner\, L. '+YEAR='1989 '+TITLE='A tutorial on Hidden Markov Models and Selected Applications in Speech Recognition '+SOURCE='PRINT'+TYPE='APA›

Fig. 3. NooJ XML output

The main syntactic grammar is built with many smaller subgraphs (month, year, page, etc.) some of which are reused at several positions making the grammar easier and faster to write and maintain (Fig. 1). Not all three reference styles use as simple and concise a grammar as the one we built for APA. They actually grew in the complexity and required some additional nodes in order to perform according to the requirements of each style (compare Figs. 1 and 2).

Our grammar has been trained on the *University of Pittsburgh* and *The Purdue Online Writing Lab* sets of data pertaining APA, MLA and Chicago citing styles. Both sites explain various ways of citing works such as books, articles and websites with examples for each of the citing styles. We finished the testing phase when our grammar reached the f-score of 1. After the parsing, the concordance window provides us with the references found but also with the XML-like output (Fig. 3) that consists of attribute = value sets. This kind of output can easily be exported and managed through other programs.

3 Building the Trees

Our website[2] uses PHP 5 powered by Apache and JavaScript with addition of free JavaScript vector library called Raphaël made by Dmitry Baranovskiy [8]. The site is divided into three main sections: homepage (basic information and user instructions), the core of the site (uploading documents and generating the tree), and the public *ReferenceTrees* section (displaying trees that the authors have made public). We will describe here the middle section, i.e. the tree structure and generation of the tree.

---

[2] URL: www.ikstudenstkiprojekti.ffzg.hr/ReferenceTrees/index.php.

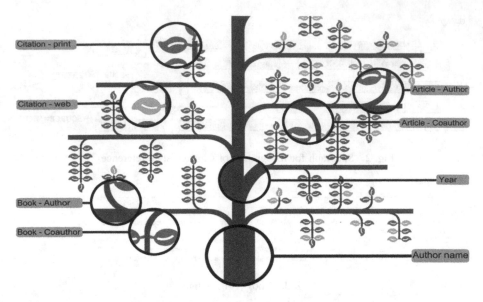

Fig. 4. Explanation of the Tree structure parts

As already noted in the introduction, we imagine our visualization as a realistic tree (Fig. 4) that gives an overview of all the books (left half of the tree) and articles (right half of the tree) written by an author. Each main branch (both sides) stands for a year when the book/article was published. On each main branch there are twigs representing a specific book/article. Twigs presenting papers written only by an ego are positioned on the upper side of the main branch, while the papers written in co-authorship are on the lower side. References used in a book/article are shown as leaves of each twig. Leaves are color coded depending on a type of a reference (book, article or a web site). The tree sections are animated giving more information when selected. If an author has used the same reference in more of his/her papers, all of the matching references are highlighted upon the selection of any one of them. Although at this point, only the individual trees may be explored, we feel that this is the first step in building and exploring author reference networks.

We can explain the procedure of building the *ReferenceTree* via PHP with the following six main steps[3]:

1. Extract data from database to JavaScript/JQuery array $dataBooks
2. Calculate height of vertical tree branch using data from array $totalHeight;
3. Draw main vertical tree branch with width of 90px and height of $totalHeight;
4. Iterate through $dataBooks and for every $year draw horizontal branch on the left;
5. Repeat step 4 but use $dataPapers and draw horizontal branches on the right;

---

[3] Due to the length and complexity of real and pseudo codes used, in this paper we are only giving the main steps while the visual demo and JavaScript source code are available at: http://www. ikstudenstkiprojekti.ffzg.hr/CitationTrees/exampleTree.php.

6. (a) For each $Reference in $dataBooks and $dataPapers draw vertical twigs on the year branch and calculate twig height;
   (b) With each previous iteration create and position leaf SVG DOM element from Raphael.js library using CSS and color it depending on the source type.

4 Conclusion and Future Work

We have presented a tree-shape *ReferenceTrees* model for visualizing bibliographies used in scientific books or articles by an author (ego). We have managed to incorporate multiple dimensions (author, year of publishing, type of publication, authorship or co-authorship, number of references, source of a reference, repeated or a unique reference among all the published works) into one relatively simple representation - tree. In this process, we have taken few steps (parsed the text, extracted the data, build the trees) and used several technologies (NooJ, XML, HTML5, JavaScript, PHP) so that we can produce as complete a tool for building reference trees as possible.

We see many opportunities in advancing our *ReferenceTrees* proposed in this paper. As our future work, we are considering the ways to incorporate the size and the shape of a leaf to show some additional characteristics to our trees (information about the scientific field of the article/book, or co-reference relations and self-citations). Also, the color and the thickness of a twig may be used to show how many times that specific publication has been cited by others in our database or the language of a publication (it would be interested to see in how many languages an author publishes). This information may be further used in placing the trees with similar structures closer to one another in a scientific forest, or the forest may switch on the lights of the trees that use references belonging to specific branch of a science e.g. linguistics or even more specific e.g. morphology. Taking into account all the possibilities our *ReferenceTrees* offer, we believe that they may find their usage in digital library catalogues, or scientific social networking sites but may also give another perspective to scientific development as a whole.

References

1. Sallaberry, A., Fu, Y.-C., Ho, H.-C., Ma, K.-L.: ContactTrees: a technique for studying personal network data. CoRR, abs/1411.0052 (2014)
2. Fung, T.-L., Ma, K.-L.: Visual characterization of personal bibliographic data using a botanical tree design. In: Electronic Proceedings of IEEE VIS 2015 Workshop on Personal Visualization: Exploring Data in Everyday Life (2015). http://www.vis4me.com/personalvis15/papers/fung.pdf
3. Fung, T.-L., Chou, J.-K., Ma, K.-L.: Comparing characteristics of majors using egocentric botanic-trees (2015). http://vacommunity.org/ieeevpg/viscontest/2015/entries/6.html
4. Sallaberry, A., Ma, K.-L.: Visualizing InfoVis Researchers with ContactTrees (2012). http://web.cse.ohio-state.edu/ ~ raghu/teaching/CSE5544/Visweek2012/infovis/posters/sallaberry.pdf

5. Sallaberry, A., Fu, Y.-C., Ho, H.-C., Ma, K.-L.: Contact trees: network visualization beyond nodes and edges. PLoS ONE **11**(1), e0146368 (2016). doi:10.1371/journal.pone.0146368
6. Chen, C., Dubin, R., Schultz, T.: Science mapping. In: Khosrow-Pour, M. (ed.) Encyclopedia of Information Science and Technology, 3rd edn. IGI Global (2014). doi:10.4018/978-1-4666-5888-2.ch410
7. Silberztein, M.: NooJ manual. http://www.nooj4nlp.net, 223 p. (2003)
8. Baranovskiy, D.: Raphaël -JavaScript Library, http://raphaeljs.com. Accessed 17 Jan 2016

Semantic Publishing Challenge: Bootstrapping a Value Chain for Scientific Data

Sahar Vahdati[1(✉)], Anastasia Dimou[4], Christoph Lange[1,2],
and Angelo Di Iorio[3]

[1] University of Bonn, Bonn, Germany
vahdati@uni-bonn.de, math.semantic.web@gmail.com
[2] Fraunhofer IAIS, Sankt Augustin, Germany
[3] Università di Bologna, Bologna, Italy
angelo.diiorio@unibo.it
[4] Ghent University – iMinds – Data Science Lab, Ghent, Belgium
anastasia.dimou@ugent.be

Abstract. The objective of the Semantic Publishing (SemPub) challenge series is to bootstrap a value chain for scientific data to enable services, such as assessing the quality of scientific output with respect to novel metrics. The key idea was to involve participants in extracting data from heterogeneous resources and producing datasets on scholarly publications, which can be exploited by the community itself. Differently from other challenges in the semantic publishing domain, whose focus is on *exploiting* semantically enriched data, SemPub focuses on *producing* Linked Open Datasets. The goal of this paper is to review both (i) the overall organization of the Challenge, and (ii) the results that the participants have produced in the first two challenges of 2014 and 2015 – in terms of data, ontological models and tools – in order to better shape future editions of the challenge, and to better serve the needs of the semantic publishing community.

1 Introduction

Semantic publishing – defined as the *use of Semantic Web technologies to make scholarly publications and data easier to discover, browse and interact with* [15] – is a lively research area in which a big number of projects and events have emerged and showcase the potential of Linked Data technology. Extracting, annotating and sharing scientific data (by which, here, we mean standalone research datasets, data inside documents, as well as metadata about datasets and documents), up to building new research on them, will lead to a *data value chain* producing value for the scientific community [10].

Bootstrapping and enabling such value chains is not easy. A solution that has proved to be successful in other communities is to run *challenges*, i.e. competitions in which participants are asked to complete tasks and have their results ranked, often in objective way, to determine the winner. Even a number of

© Springer International Publishing AG 2016
A. González-Beltrán et al. (Eds.): SAVE-SD 2016, LNCS 9792, pp. 73–89, 2016.
DOI: 10.1007/978-3-319-53637-8_9

projects have been launched to accelerate this process, for instance LinkedUp[1] or Apps for Europe[2]. The success of the LAK[3] or Linked Up[4] Challenges is worth mentioning here. However, these challenges focus on *exploiting* scholarly linked data for different purposes (for instance, to monitor progress) but less on actually *producing* such datasets.

To this end, we started a series of Semantic Publishing Challenges (SemPub), aiming at the production of datasets on scholarly publications. To the best of our knowledge, this was the first challenge of its kind. Now, in 2016, we are running the 3rd edition of SemPub and we believe it is good time to review the challenge and share some lessons learned with the community. On the other hand, community feedback can help us shape the future of SemPub. Continuous refinement is in fact a key aspect of our vision.

Section 2 of this paper reviews the background of SemPub, its history, structure and evaluation methods. Section 3 presents lessons learned from the challenge organization and Sect. 4 from the overall approaches of the submitted solutions. Section 5 concludes and provides an outlook to how future SemPub challenges will take these lessons into account.

2   History of the SemPub Challenge

We draw a brief history of the SemPub Challenge to provide the necessary background for the following discussion. More detailed reports have been published separately for the 2014 [8] and 2015 [1] challenges.

We started in 2014, reasoning about a challenge in the semantic publishing domain that could be measured in an objective way. This was difficult because of the tension between finding appealing and novel tasks and measuring them. We thus asked participants to extract data from scholarly papers and to produce an RDF dataset that could be used to answer some relevant queries: concretely, queries about the *quality* of scientific output. We were aware of other topics of interest for the community – nanopublications, research objects, etc. – but focused on papers only to bootstrap the initiative and to start collaboratively producing initial data.

We designed different tasks, sharing the same organization, rules and evaluation procedure. For each task, we published a set of queries in natural language and asked participants to translate them into SPARQL and to submit a dataset on top of which these queries would run. In line with the general rules for the new Semantic Web Evaluation Challenge track at ESWC, we also published a training dataset (TD) on which the participants could test and train their extraction tools. A few days before the submission deadline, we published an evaluation dataset (ED): the input for the final evaluation.

---

[1] http://linkedup-project.eu/.

[2] http://www.appsforeurope.eu/.

[3] Learning Analytics and Knowledge; see http://meco.l3s.uni-hannover.de:9080/wp2/?page_id=18.

[4] http://linkedup-challenge.org/.

The evaluation consisted of comparing the output of these queries, given as CSV, against a gold standard and measuring precision and recall. All three editions used the same evaluation procedure, but the tasks were refined over time. Table 1 summarizes all tasks, their data sources and queries.

Table 1. Description, source and format of the tasks in SemPub editions (2014–2016).

	2014	2015	2016
Task1	Extracting data on workshops' quality indicators; Source: CEUR-WS.org		
Format: HTML+PDF	Format: HTML	Format: HTML	
Task2	Extracting data on citations Source: PubMed Format: XML	Extracting data on affiliations, citations, funding Source: CEUR-WS.org Format: PDF	Extracting data on affiliations, internal structure, fundings Source: CEUR-WS.org Format: PDF
Task3	Open tasks: showcase semantic publishing applications	Interlinking Sources: CEUR-WS.org, Colinda, DBLP, Lancet, Semantic Web Dog Food (SWDF), Springer LD)	Interlinking Sources: CEUR-WS, Colinda, DBLP, Springer LD

Two tasks have been defined at the very beginning (see [8] for full details and statistics):

– **Task 1:** participants were asked to extract information from selected CEUR-WS.org[5] workshop proceedings volumes (HTML tables of content using different levels of semantic markup, plus PDF full text) to enable the computation of indicators for the workshops' quality assessment. They were asked to answer 20 different queries.
– **Task 2:** participants were asked to extract data about citations, to enable precise assessment of linking, sharing and evaluating research through citations. The dataset included a set of XML-encoded research papers, taken from PubMedCentral and Pensoft Open Access archives, and heterogeneous in terms of internal structure, styles and numbers. Both dataset and queries were completely disjoint from Task 1.

Having called for submissions, we received feedback from the community that mere information extraction, even if motivated by quality assessment, was not the most exciting task related to the future of scholarly publishing, as it assumed

---

[5] http://ceur-ws.org/.

a traditional publishing model. Furthermore, to address the primary target of the challenge, i.e. "publishing" rather than just "metadata extraction", we widened the scope by adding an *open task*, whose participants were asked to showcase data-driven applications that would eventually support publishing. We received a good number of submissions; winners were selected by a jury.

In 2015 we were asked to include only tasks that could be evaluated in a fully objective manner, and thus discarded the open task. We reduced the distance between Tasks 1 and Task 2 by using the same dataset for both. We transformed Task 2 into a PDF mining task and thus moved all PDF-related queries there. The rationale was to differentiate tasks on the basis of the competencies and tools required to solve them, but to make tasks interplay on the same dataset.

CEUR-WS.org data has become the central focus of the whole Challenge, for two reasons: on the one hand, the data provider (CEUR-WS.org) takes advantage of a broader community that builds on its data, which, before the SemPub Challenges, had not been available as linked data. On the other hand, data consumers gain the opportunity to assess the quality of scientific venues by taking a deeper look into their history, as well as the quality of the publications. While Task 1 queries remained largely stable from 2014 to 2015, the queries for Task 2 changed, mainly because the setting was completely new (PDF rather than XML sources), and we wanted to explore participants' interest and available solutions. We asked them to extract data not only on citations but also on affiliations and fundings.

In 2015 we added a new Task 3, focusing on interlinking the dataset the winners of the first Challenge had extracted from a single source to further relevant datasets. Participants had to make such links explicit and exploit them to answer comprehensive queries about events and persons. CEUR-WS.org on its own provides incomplete information about conferences and persons, which can be complemented by interlinking with other datasets to broaden the context and to allow for more reliable conclusions about the quality of scientific events and the qualification of researchers.

Continuity is the key aspect of 2016 edition. The tasks are unchanged (allowing previous participants to use and refine their tools), except for details: Task 2, in particular, is extended to structural components of papers and does not include citations anymore.

3   Lessons Learned from the Challenge Organization

In this section we discuss lessons learned from our experience in organizing the challenge. Our goal is to distill some generic guidelines that could be applied to similar events, starting from the identification of critical issues, errors and strengths of our initiative. The primary focus of the paper is on (even unexpected) aspects that emerged while running the challenge. This section will also present the lessons learned by looking at the solutions and data produced by the participants. We have grouped the lessons in four categories for clarity, even though there is some overlap between them.

3.1   Lessons Learned on Defining Tasks

The definition of the tasks is the most critical part of organizing a challenge. In our case, it was difficult to define appealing tasks that bridge the gap between building up initial datasets and exploring possibilities for innovative semantic publishing. As discussed in Sect. 2, we refined the tasks over the years according to the participants' and organizers' feedback. Overall, we think that tasks could have been improved in some parts – and undeniably other interesting ones could have been defined – but they were successful. There are other less evident issues which are worth discussing here.

L1.1. Continuity: allow users to re-submit the improved version of their tool over different editions. One of the goals of the first edition of the challenge was also to explore the interest of the participants. Exploiting such feedback and creating a direct link between different editions is a success key factor. In 2015, in fact, the Challenge was re-organized aiming to commit participants to re-submit overall improved versions of their first year submissions. Results were very good, as the majority of first year's participants competed for the second year too. Continuity is also a key aspect of SemPub2016, whose tasks are the same as last year's edition, allowing participants to reuse their tools to adapt to the new call after some tuning.

L1.2. Split tasks with a clear distinction of the competencies required to complete them. One of the main problems we faced was that some tasks were too difficult. In particular the Task 2 – extraction from XML and PDF – showed unexpectedly low performance. The main reason, in our opinion, is that the task was actually composed of two sub-tasks that required very different tools and technologies: some queries required participants to basically map data from XML/PDF to RDF, while the others required additional processing on the content. Some people were discouraged to participate as they only felt competitive for the one and not for the other. Our initial goal was to explore a larger amount of information and to give participants more options but, in retrospect, such heterogeneity was a limitation. A sharper distinction between tasks would have been more appropriate. In particular, it is important to separate tasks on plain data extraction from those on natural language processing and semantic analysis.

L1.3. Involve participants in advance in the task definition. Though we collected some feedback when designing the tasks, we noticed that such preliminary phase was not given enough relevance. The participants' early feedback can help to identify practical needs of researchers and to shape tasks. Talking with participants, in fact, we envisioned alternative tasks, such as finding high-profile venues for publishing a work, summarizing publications, or helping early career researchers to find relevant papers. Proposing tasks emerged from the community can be a winning incentive to participate.

3.2   Lessons Learned on Building Input Datasets

The continuity between tasks (L1.1) can be applied to the datasets as well:

L2.1. Use the same data source for multiple editions. We noticed benefits of using the same data sources across multiple editions of the Challenge. From the task 1 of the 2014 edition, in fact, we obtained an RDF dataset that served as the foundation to build the same task in 2015 and 2016. Participants were able to reuse their existing tools and to extend the previously-created knowledge-bases with limited effort. For the other tasks, which were not equally stable, we had to *rebuild the competition* every year without being able to exploit the past experience.

L2.2. Design all three tasks around the same dataset. Similarly, it is valuable to use the same dataset for multiple tasks. First of all, for the participants: they could extend their existing tools to compete for different tasks, with a quite limited effort. This also opens new perspectives for future collaboration: participants' work could be extended and integrated in a shared effort for producing useful data. It is also worth highlighting the importance of such uniformity for the organizers. It reduces the time needed to prepare and validate data, as well as the risk of errors and imperfections. Last but not least, it enables designing interconnected tasks and producing richer output.

L2.3. Provide an exhaustive description of the expected output on the training dataset. An aspect that we underestimated in the first editions of the Challenge was the description of the training dataset. While we completely listed all papers we did not provide enough information on the expected output: we went into details for the most relevant and critical examples but we did not provide the exact expected output for all papers in the training dataset. Such information should instead be provided as it impacts directly the quality of the submissions and help participants to refine their tools.

3.3   Lessons Learned on Evaluating Results

All three editions of the Challenge shared the same evaluation procedure (see Sect. 2 for more details). The workflow presented some weaknesses, especially in the first two years, which we subsequently addressed. Three main guidelines can be derived from these issues.

L3.1. Consider all papers in the final evaluation. Even though we asked participants to run their tools on the whole evaluation dataset, we considered only some exemplary papers for the final evaluation. These papers have been randomly selected from clusters representing different cases, which participants were required to address. Since these papers were representative of these cases we received a fair indication of the capabilities of each tool. On the other hand, some participants were penalized as their tool could have worked well on other values, which were not taken into account for the evaluation. In the third edition, we will radically increase the coverage of the evaluation queries and their number in order to assure that greatest part of the dataset (or the whole dataset) is covered.

L3.2. Make evaluation tool available during the training phase. The evaluation was totally transparent and all participants received detailed feedback about their scores, together with links to the open source tool used for the final evaluation. However we were able to release the tool only after the Challenge. It is instead more helpful to make it available during the training phase, as participants can refine their tool and improve the overall quality of the output. Such an approach reduces the (negative) impact of output imperfections. Though the content under evaluation was normalized and minor differences were not considered as errors, some imperfections were not expected and were not handled in advance. Some participants, for instance, produced CSV files with columns in a different order or with minor differences in the IRI structure. These all could have been avoided if participants received feedback during the training phase, with the evaluation tool available as a downloadable stand-alone application or as a service.

L3.3. Use disjoint training and evaluation datasets. A 2015 participant raised the issue that we underestimated when designing the evaluation process: the evaluation dataset was a superset of the training one. This resulted in some over-training of the tools, and caused imbalance in the evaluation. It is more appropriate to use completely disjoint datasets, a solution we are implementing for the last edition.

3.4 Lessons Learned on Expected Output and Organizational Aspects

Further suggestions can also be derived from the Challenge's organizational aspects, in particular regarding the expected outcome:

L4.1. Define clearly the license of produced output. Some attention should be given to the licensing of the output produced by the participants. We did not explicitly say which license they should use: we just required them to use an open license on data (at least permissive as the source of data) and we encouraged open-source licenses on the tools (but not mandatory). Most of the participants did not declare which exact license applies to their data. This is an obstacle for the reusability: especially when data come from heterogeneous sources and are heterogeneous in content and format, as in the case of CEUR-WS papers, it is very important to provide an explicit representation of the licensing information.

L4.2. Define clearly how the output of the challenge will be used. The previous observation can be generalized into a wider guideline about reusability. It is in fact critical to state how the results of the challenge will be eventually used, in order to encourage and motivate participants. The basic idea of the Challenge was to identify the best performing tool on a limited number of papers and to use the winning tool – or a refined version – to extract the same data on the whole CEUR-WS corpus[6]. The production of the CEUR-WS

---

[6] At least, on the subset of CEUR-WS.org whose license scheme allowed us to republish metadata.

Linked Open Dataset was actually slower than expected and we are finalizing it in these days. This is a critical issue: participants' work should not target the challenge only, but it should produce an output that is directly reusable by the community.

L4.3. Study conflicts and synergies with other events. The last guideline is not surprising and was confirmed by our experience as well. In 2015, in fact, we introduced a task on interlinking. The community has been studying interlinking for many years and a lot of research groups could have participated in the task (and produced very good results). However we did not receive enough submissions. One of the issues – not the only one, communication might be another – is the conflict with events like OAEI (Ontology Alignment Evaluation Initiative). Even though Task 3 of SemPub2015 did not intend to cover the specialized scope of OAEI, but rather put the interlinking task in a certain use case scope that merely serves in aligning the tasks output among each other and with the rest LOD cloud. The study of overlapping and similar events should always be kept in mind. Not only to identify potential conflicts but also to generate interest: the fact that the SePublica workshop was at ESWC 2014, for instance, was positive since we had fruitful discussions with the participants and the two events could benefit each other.

4   Lessons Learned from Submitted Solutions

In this section we discuss lessons learned from the participants' solution. We start with an overview of the solutions; next, we group the lessons into four categories: lessons on submitted tools, used ontologies, submitted data and evaluation process; even though there is some overlap between these aspects.

4.1   Solutions by Task

Task 1 Solutions – 2015 and 2014. There were four different solutions proposed to address Task 1 in 2014 and 2015. Three participated in both editions, whereas the fourth solution participated only in the second edition.

Solution 1.1. [5,6] presented a case-specific crawling based approach for addressing Task 1. It relies on an extensible template-dependent crawler that uses sets of special predefined templates based on XPath and regular expressions to extract the content from HTML and to convert it in RDF. The RDF is then processed to merge resources using fuzzy-matching. The use of the crawler turns the system tolerant to invalid HTML pages. This solution improved its precision in 2015 as well the richness of the data model.

Solution 1.2. [2,3] exploited a generic tool for generating RDF data from heterogeneous data. It uses the RDF Mapping Language (RML)[7] to define how data extracted from CEUR-WS Web pages should be semantically annotated. RML extends R2RML to express mapping rules from heterogeneous data to

---

[7] http://rml.io.

RDF. CSS3 selectors are considered to extract the data from the HTML pages. The RML mapping rules are parsed and executed by the RML Processor[8]. In 2015 the solution reconsidered its data model and was extended to validate both the mapping documents and the final RDF, resulting in an overall improved quality dataset.

Solution 1.3. [12,13] designed a case-specific solution based on a linguistic and structural analyzer. It uses a pipeline based on the GATE Text Engineering Framework. To produce annotations, it relies on chunk-based and sentence-based support vector machine (SVM) classifiers which are trained using the CEUR-WS proceedings with microformat annotations. The annotation sanitizer has a set of heuristics which are applied to fix imperfections and interlink annotations. The produced dataset is also extended with information retrieved from external resources.

Solution 1.4. [9] presented an application of the FITLayout framework[9] This solution participated in the Semantic Publishing Challenge only in 2015. It combines different page analysis methods, i.e. layout analysis and visual and textual feature classification to analyze the rendered pages, rather than their code. The solution is quite generic but needs to be domain/case-specific at certain phases (model building step).

All solutions are summarised in Table 2, which also add details about the languages and technologies exploited by the participants.

Table 2. HTML-code-based and content-based solutions for Task 1.

	Solution 1	Solution 2	Solution 3	Solution 4
Primary method	Crawling	Generic solution for abstracted mappings	Linguistic and structural analysis	Visual layout Multi-aspect content analysis
Case-specific	YES	NO	YES	NO
Template-based	YES	YES	NO	~NO
Implementation basis	–	RML tools	–	FITLayout
Implementation Language	Python	Java	–	Java/HTML
Mappings/Rules	XPath (embedded in the code)	–	RML/CSS (abstracted from the code)	Hard coded
RegEx	YES	YES	–	YES

Task 2 Solutions – 2015. Solution 2.1. CERMINE [16] is an open source system for extracting structured metadata and references from scientific publications published as PDF files. It has a loosely captured architecture and a modular

---

[8] https://github.com/RMLio/RML-Mapper.
[9] http://www.fit.vutbr.cz/~burgetr/FITLayout/.

workflow which is based on supervised and unsupervised machine-learning techniques which simplifies the system's adoption to new document layouts and styles.

Solution 2.2. [4] implemented a processing pipeline that analyzes the structure of a PDF document incorporating a diverse set of machine learning techniques, unsupervised to extract text blocks and supervised to classify blocks into different meta-data categories. Heuristic are applied to detect the reference section and sequence classification to categorize the tokens of individual references strings. Finally, named entity recognition (NER) are used to extract references to grants, funding agencies and EU projects.

Solution 2.3. [11] presented Metadata And Citations Jailbreaker (MACJa – IPA), a tool that integrates hybrid techniques based on Natural Language Processing (NLP) and incorporating FRED, a novel machine reader. It also includes modules to query external services to enhance and validate data.

Solution 2.4. [14] presented a system composed by two modules: a text mining pipeline based on the GATE framework to extract structural and semantic entities, leveraging also existing NER tools, and a LOD exporter, to translate the document annotations into RDF according to custom rules.

Solution 2.5. [7] relies on a rule-based and pattern matching approach, implemented in Python and some external services for improving the quality of the results (for instance, DBLP for validating author's data). It also relies on an external tool to extract the plain text from PDFs.

Solution 2.6. [12] extended their framework used for Task 1 (and indicated as Solution 1.3 before) to extract data from PDF as well. Their pipeline includes text processing and entity recognition modules and employs external services for mining PDF articles. Table 3 represents tools and its components:

4.2   Lessons Learned from the Tools

L5.1. There are both generic and ad hoc solutions. All solutions were methodologically different among each other. For Task 1, for instance, two solutions (1.1 and 1.3) primarily consisted of a tool developed specific to this task, whereas the other two solutions wrote task-specific templates in the otherwise generic implementations (adaptive to other domains). In the later case, Solution 1.2 abstracted the case-specific aspects from the implementation, whereas Solution 1.4 kept them inline with the implementation. It becomes, therefore, clear that there are alternative approaches which can be used to produce RDF datasets.

L5.2. There are HTML code and content-based approaches to information extraction. Even though solutions were methodologically different, two main approaches for dealing with the HTML pages prevailed: HTML-code-based and content-based.

Table 3. Tools and its components.

	Solution 2.1	Solution 2.2	Solution 2.3	Solution 2.4	Solution 2.5	Solution 2.6
Implementation language	Java/HTML	Java	Java/Python	Java	Python/HTML	Java
Implementation on basis	CERMINE (https://github.com/CeON/CERMINE)	code-annotator (http://code-annotator.know-center.at)				
Components	LibSVM (https://www.csie.ntu.edu.tw/~cjlin/libsvm/), GRMM (http://mallet.cs.umass.edu/grmm/), Mallet (http://mallet.cs.umass.edu/)	OpenNLP (https://opennlp.apache.org/), GATE (https://gate.ac.uk/), ParsCit (http://wing.comp.nus.edu.sg/parsCit/), crfsuite (http://www.chokkan.org/software/crfsuite/)	FRED (http://wit.istc.cnr.it/stlab-tools/fred) CrossRef API (http://api.crossref.org/), FreeCite (http://freecite.library.brown.edu/), Stanford CoreNLP (http://stanfordnlp.github.io/CoreNLP/), NLTK (http://www.nltk.org/), WordNet (https://wordnet.princeton.edu/), BabelNet (http://babelnet.org/))	GATE (https://gate.ac.uk/)	Grab spider (http://grablib.org/), BeautifulSoup (http://www.crummy.com/software/BeautifulSoup/)	GATE (https://gate.ac.uk/), Poppler (http://poppler.freedesktop.org/), CrossRef API (http://api.crossref.org/), FreeCite (http://freecite.library.brown.edu/), Bibsonomy API (http://www.bibsonomy.org/help/doc/api.html)
PDF/character extraction	iText (http://itextpdf.com/)	Apache PDFBox (https://pdfbox.apache.org/)	PDFMiner (http://www.unixuser.org/~euske/python/pdfminer/)	Xpdf (http://foolabs.com/xpdf/)	PDFMiner	PDFX (pdfx.cs.man.ac.uk)
Open Source	NO	YES/broken	NO	NO	YES	NO
Intermediate representation	XML/TrueViz, NLM JATS/XML		JSON		TXT, HTML	
Ontologies reused (We abbreviate well-known ontologies by their prefixes according to prefix.cc http://prefix.cc.)		dul, dbpedia, schema	SPAR ontologies	http://lod.semanticsoftware.info/pubo/ deo, sro	bibo, foaf, dc, swrc, dbpedia, arpfo	swrc, biro, fabio, pro

4.3   Lessons Learned from Models and Ontologies

L6.1. All solutions used almost the same data model (Task 1). All solutions of Task 1 tend to converge regarding the model of the data. The same occurs but on a higher level in the case of Task 2. In particular for Task 1, Solution 1.4 domain modeling was inspired by the model used in Solution 1.1, with some simplifications. Note also that Solution 1.2 was the winner solution in 2014. Based on the aforementioned, we observe a trend of converging regarding the model the CEUR-WS data set should have, as most of the solutions converge on the main identified concepts in the data (Conference, Workshop, Proceedings, Paper and Person).

   L6.2. All solutions used almost the same vocabularies for the same data (Task 1). There is a wide range of vocabularies and ontologies that can be used to annotate scholarly data. However, most of the solutions preferred to (re)use almost the same existing ontologies and vocabularies (see Table 4 for Task 1). This is a good evidence that the spirit of vocabulary reuse gains traction. However, it is interesting that different solutions used the same ontologies to annotate the same data differently (see L6.3).

Table 4. Vocabularies for the same data for Task 1.

Vocabulary (We abbreviate well-known ontologies by their prefixes according to prefix.cc http://prefix.cc.)	Solution 1	Solution 2	Solution 3	Solution 4
bibo	✓	✓	–	✓
biro	–	–	✓	–
co	–	–	✓	–
dbpedia	✓	Java	✓	✓
dc	✓	Java	✓	✓
dcterms	✓	✓	–	✓
event	–	✓	–	–
fabio	–	✓	✓	–
foaf	✓	✓	✓	✓
frbr	–	–	✓	–
pro	–	–	✓	–
skos	✓	–	–	–
swc	✓	–	–	✓
swrc	✓	✓	✓	✓
timeline	✓	–	–	✓
others/custom	–	–	✓	✓

Table 5. Vocabularies for different annotations for Task 1.

Task 1/2015	Solution 1	Solution 2	Solution 3	Solution 4
Person	foaf:Person	foaf:Person	foaf:Person	foaf:Person
Paper	bibo:Article	swrc:InProceedings	swrc:Publication	swc:Paper
Conference	swc:OrganizedEvent	bibo:Conference	swrc:Conference	swc:ConferenceEvent
Proceeding	bibo:Proceeding	bibo:Proceedings	swrc:Proceedings	swc:Proceedings
Proceeding	bibo:Workshop	bibo:Workshop	swrc:Workshop	swc:section

L6.3. Different solutions used different annotations (Task 1). Even though all solutions used almost the same vocabularies, not all solutions used the same classes to annotate same entities. To be more precise, all solutions only converged on annotating persons using *foaf:Person*. For the other main concepts the situation was heterogeneous, as reported in Table 5.

L6.4. Different solutions used different vocabularies for the same data (Task 2). In contrast to Task 1 solutions which all intuitively converged on using the same vocabularies and ontologies, Task 2 solutions use relatively different vocabularies and ontologies, but again pre-existing ones, to annotate same entities appearing in the same data. However, most of the Task 2 solutions use sub-ontologies of the family of SPAR ontologies. It is interesting to observe if the Task 2 solutions of 2016 will converge towards using same ontologies, being inspired one from the other, or if solutions will keep using different vocabularies.

4.4   Lessons Learned from Submitted RDF

L7.1. Overall dataset improved over successive challenges. From the first edition to the second edition of the Semantic Publishing Challenge, we expected that participants who re-submit their solutions would have improve the overall dataset, rather than optimize it for answering the queries. All three solutions of Task 1 both in 2014 and 2015 edition modified the way they represented their model in 2014 for their submissions in 2015 which resulted in corresponding improvements to the overall dataset. Although this happens to a certain extend and indeed the results were more satisfying, we still see that there is room for overall improvement.

L7.2. Participants preferred custom solutions Custom solutions for a particular task, such as publishing CEUR-WS.org proceedings, may obviously result in more accurate output in terms of answers to queries, however they lack repurposeability, as they cannot be reused for other input data. Moreover, despite the fact that there are generic tools for extracting RDF datasets, challenge participants preferred to develop custom solutions. This can be interpreted as a lack of alternatives of HTML specific tools to address the task.

L7.3. Striking differences in coverage We further observed that solutions rarely agree upon the extracted information. Overall, we observe significant differences in respect to the number of identified entities per category. The results for Task 1 are summarized in the Table 6.

Table 6. The number of instances produced for each class (for Task 1).

–	Solution 1.1	Solution 1.2	Solution 1.3	Solution 1.4
#Conferences	46	46	51	47
#Workshops	252	1393	127	198
#Proceedings	243	1392	202	1353
#Papers	3801	2452	720	2470
#Persons	6700	6414	3402	11034

Let us consider the proceedings for example. Apparently, Solution 1.1 and Solution 1.3 used the individual pages to identify the proceedings, whereas Solution 2 and Solution 4 used the index page to identify the proceedings, this is the reason that there is so big difference in the numbers. The number of identified papers is also significantly different among the different solutions, but in the case of persons we observe the most variation in terms of numbers. However, the more the solutions improve, the more we expect to find solutions that converge at least regarding the number of retrieved and/or distinctly identified entities.

L7.4. RDF datasets differed significantly w.r.t. statistics. Produced datasets were also very heterogeneous in term of size, number of triples, entities and so on. Table 7 summarizes the statistics for Task 1.

Table 7. Statistics about the produced dataset (for Task 1).

–	Solution 1.1	Solution 1.2	Solution 1.3	Solution 1.4
Dataset size	9.6 M	6.6 M	3.8 M	5.1 M
#triples	177,752	95,015	62,231	79,444
#entities	11,208	11,719	11,589	19,090
#properties	46	23	42	23
#classes	10	10	10	6

Note that the size of the largest dataset is almost double the size (9.6 M) of the smallest (5.1 M). Similarly, the largest dataset in terms of triples (~180,000), contains three times more triples compared to the smallest (~63,000) to model the same data set. Solution 4 is the only one which required significant larger number of entities to represent the same data. Considering that Solution 4 presents a very large number of persons, the correspondingly high number of entities is not so surprising.

L7.5. No provenance or other metadata. Unfortunately, no team intuitively provided any provenance or other metadata information. In particular licensing metadata information is of crucial importance for subsequent use of datasets.

4.5 Lessons Learned from the Solutions with Respect to the Evaluation

L8.1. Performance ranking of the tools evolved but not as expected. In 2015 the performance ranking of the three tools evolved from 2014 has not changed but their performance has improved except for Kolchin et al., who improved precision but not recall. Disregarding the two queries that were new in 2015, the tool by Kolchin et al., which had won the best performance award in 2014, performs almost as well as Milicka's/Burget's.

L8.2. New and legacy solutions were both valuable. Task 1 participants both in 2014 and 2015 had an improved version of different aspects of their solution which resulted in correspondingly improved versions of the final dataset. The new solution which introduced a fundamentally new approach and participated in Task 1 achieved equally good results as the best solution of 2014. In conclusion, legacy solutions might be able to improve and bring stable and good results, however there is still room for improvement and mainly for fundamentally new ideas that surpass problems that legacy solutions can not deal with.

L8.3. Newly introduced approaches have equal chances in winning the challenge. The winners of Task1 in 2014 participated in 2015 with an improved version of their tool but they did not win. The 2015 winner was a new tool with a brand new approach. The winners were not the same in the two versions of the challenge, creativity won.

5 Conclusions

One of the objectives of the SemPub Challenge series is to produce Linked Data that contribute to improving the scholarly communication. This includes supporting researchers finding relevant and high-quality papers by exploiting the information available in these datasets. Semantic Web technologies in this context do not only solve isolated problems, but generates further value in that data can be shared, linked to each other, and reasoned on.

The goal of this work is to shade light on the first editions of the Challenge and to distill some lessons learned from our experience. In particular, we were interested in both organizational aspects and evidences from the solutions proposed by the participants. Our conclusion is that we are moving in the right direction but the goal has not been fully reached yet. There are several positive aspects, among which the high participation and the quality of the produced results. The possibility of sharing knowledge and solutions among participants was another key factor of SemPub. The Challenge allowed us to share experience on semantifying scholarly data, using some ontological models, refining and extending existing datasets. On the other hand, our analysis showed that some other aspects have to be necessarily improved. In particular, we have to make the produced output well integrated in the Linked Open Data ecosystem and exploited by the community.

The next step in fact is to investigate what are the services that we can build on top of the produced data and how they can be offered. Some natural (and challenging) questions arise: what services can already be delivered based on the data we currently have? How do we need to extend these data to provide novel services? What would be the interface of such services look like? Which functionalities should be implemented first? The challenge will also be to turn all these questions into new material for a new Challenge, even better if measurable in an objective way.

Acknowledgments. We would like to thank our peer reviewers for their constructive feedback. This work has been partially funded by the European Commission under grant agreement no. 643410.

References

1. Di Iorio, A., Lange, C., Dimou, A., Vahdati, S.: Semantic publishing challenge - assessing the quality of scientific output by information extraction and interlinking. In: Gandon, F., Cabrio, E., Stankovic, M., Zimmermann, A. (eds.) SemWebEval 2015. CCIS, vol. 548, pp. 65–80. Springer, Cham (2015)
2. Dimou, Anastasia, Sande, Miel, Colpaert, Pieter, Vocht, Laurens, Verborgh, Ruben, Mannens, Erik, Walle, Rik: Extraction and Semantic Annotation of Workshop Proceedings in HTML Using RML. In: Presutti, Valentina, Stankovic, Milan, Cambria, Erik, Cantador, Iván, Iorio, Angelo, Noia, Tommaso, Lange, Christoph, Reforgiato Recupero, Diego, Tordai, Anna (eds.) SemWebEval 2014. CCIS, vol. 475, pp. 114–119. Springer, Cham (2014). doi:10.1007/978-3-319-12024-9_15
3. Heyvaert, Pieter, Dimou, Anastasia, Verborgh, Ruben, Mannens, Erik, Walle, Rik: Semantically Annotating CEUR-WS Workshop Proceedings with RML. In: Gandon, Fabien, Cabrio, Elena, Stankovic, Milan, Zimmermann, Antoine (eds.) SemWebEval 2015. CCIS, vol. 548, pp. 165–176. Springer, Cham (2015). doi:10.1007/978-3-319-25518-7_14
4. Klampfl, Stefan, Kern, Roman: Machine Learning Techniques for Automatically Extracting Contextual Information from Scientific Publications. In: Gandon, Fabien, Cabrio, Elena, Stankovic, Milan, Zimmermann, Antoine (eds.) SemWebEval 2015. CCIS, vol. 548, pp. 105–116. Springer, Cham (2015). doi:10.1007/978-3-319-25518-7_9
5. Kolchin, Maxim, Cherny, Eugene, Kozlov, Fedor, Shipilo, Alexander, Kovriguina, Liubov: CEUR-WS-LOD: In: Gandon, Fabien, Cabrio, Elena, Stankovic, Milan, Zimmermann, Antoine (eds.) SemWebEval 2015. CCIS, vol. 548, pp. 142–152. Springer, Cham (2015). doi:10.1007/978-3-319-25518-7_12
6. Kolchin, Maxim, Kozlov, Fedor: A Template-Based Information Extraction from Web Sites with Unstable Markup. In: Presutti, Valentina, Stankovic, Milan, Cambria, Erik, Cantador, Iván, Iorio, Angelo, Noia, Tommaso, Lange, Christoph, Reforgiato Recupero, Diego, Tordai, Anna (eds.) SemWebEval 2014. CCIS, vol. 475, pp. 89–94. Springer, Cham (2014). doi:10.1007/978-3-319-12024-9_11
7. Kovriguina, Liubov, Shipilo, Alexander, Kozlov, Fedor, Kolchin, Maxim, Cherny, Eugene: Metadata Extraction from Conference Proceedings Using Template-Based Approach. In: Gandon, Fabien, Cabrio, Elena, Stankovic, Milan, Zimmermann, Antoine (eds.) SemWebEval 2015. CCIS, vol. 548, pp. 153–164. Springer, Cham (2015). doi:10.1007/978-3-319-25518-7_13

8. Lange, Christoph, Iorio, Angelo: Semantic Publishing Challenge – Assessing the Quality of Scientific Output. In: Presutti, Valentina, Stankovic, Milan, Cambria, Erik, Cantador, Iván, Iorio, Angelo, Noia, Tommaso, Lange, Christoph, Reforgiato Recupero, Diego, Tordai, Anna (eds.) SemWebEval 2014. CCIS, vol. 475, pp. 61–76. Springer, Cham (2014). doi:10.1007/978-3-319-12024-9_8

9. Milicka, Martin, Burget, Radek: Information Extraction from Web Sources Based on Multi-aspect Content Analysis. In: Gandon, Fabien, Cabrio, Elena, Stankovic, Milan, Zimmermann, Antoine (eds.) SemWebEval 2015. CCIS, vol. 548, pp. 81–92. Springer, Cham (2015). doi:10.1007/978-3-319-25518-7_7

10. Miller, G., Mork, P.: From Data to Decisions: A Value Chain for Big Data. Spart IT, ITPro (2013)

11. Nuzzolese, Andrea Giovanni, Peroni, Silvio, Recupero, Diego Reforgiato: MACJa: Metadata and Citations Jailbreaker. In: Gandon, Fabien, Cabrio, Elena, Stankovic, Milan, Zimmermann, Antoine (eds.) SemWebEval 2015. CCIS, vol. 548, pp. 117–128. Springer, Cham (2015). doi:10.1007/978-3-319-25518-7_10

12. Ronzano, Francesco, Fisas, Beatriz, Bosque, Gerard Casamayor, Saggion, Horacio: On the Automated Generation of Scholarly Publishing Linked Datasets: The Case of CEUR-WS Proceedings. In: Gandon, Fabien, Cabrio, Elena, Stankovic, Milan, Zimmermann, Antoine (eds.) SemWebEval 2015. CCIS, vol. 548, pp. 177–188. Springer, Cham (2015). doi:10.1007/978-3-319-25518-7_15

13. Ronzano, Francesco, Bosque, Gerard Casamayor, Saggion, Horacio: Semantify CEUR-WS Proceedings: Towards the Automatic Generation of Highly Descriptive Scholarly Publishing Linked Datasets. In: Presutti, Valentina, Stankovic, Milan, Cambria, Erik, Cantador, Iván, Iorio, Angelo, Noia, Tommaso, Lange, Christoph, Reforgiato Recupero, Diego, Tordai, Anna (eds.) SemWebEval 2014. CCIS, vol. 475, pp. 83–88. Springer, Cham (2014). doi:10.1007/978-3-319-12024-9_10

14. Sateli, B., Witte, R.: Automatic construction of a semantic knowledge base from CEUR workshop proceedings. In: Gandon, F., Cabrio, E., Stankovic, M., Zimmermann, A. (eds.) SemWebEval 2015. CCIS, vol. 548, pp. 129–141. Springer, Cham (2015). doi:10.1007/978-3-319-25518-7_11

15. Shotton, D.: Publishing: Open citations. Nature 502(7471) (2013)

16. Tkaczyk, D., Bolikowski, Ł.: Extracting contextual information from scientific literature using CERMINE system. In: Gandon, F., Cabrio, E., Stankovic, M., Zimmermann, A. (eds.) SemWebEval 2015. CCIS, vol. 548, pp. 93–104. Springer, Cham (2015). doi:10.1007/978-3-319-25518-7_8

Analysing Structured Scholarly Data Embedded in Web Pages

Pracheta Sahoo[2], Ujwal Gadiraju[1], Ran Yu[1],
Sriparna Saha[2], and Stefan Dietze[1(✉)]

[1] L3S Research Center, Leibniz Universität Hannover, Hannover, Germany
{gadiraju,yu,dietze}@L3S.de
[2] Indian Institute of Technology, Patna, India
{pracheta.mtmc14,sriparna}@iitp.ac.in

Abstract. Web pages increasingly embed structured data in the form of microdata, microformats and RDFa. Through efforts such as schema.org, such embedded markup have become prevalent, with current studies estimating an adoption by about 26% of all web pages. Similar to the early adoption of Linked Data principles by publishers, libraries and other providers of bibliographic data, such organisations have been among the early adopters, providing an unprecedented source of structured data about scholarly works. Such data, however, is fundamentally different from traditional Linked Data, by being very sparsely linked and consisting of a large amount of coreferences and redundant statements. So far, the scale and nature of embedded scholarly data on the Web has not been investigated. In this work, we provide a study on embedded scholarly data to answer research questions about the depth, syntactic and semantic characteristics and distribution of extracted data, thereby investigating challenges and opportunities for using embedded data as a structured knowledge graph of scholarly information.

Keywords: Linked Data · Scholarly articles · Web Data Commons · Analysis

1 Introduction

Bibliographic data is widespread on the Web. Libraries and publishers have in particular embraced the Linked Data principles and associated W3C standards throughout the past decade, making large amounts of bibliographic metadata available on the Web [2]. However, uptake and reuse is still hindered by a variety of issues, including the lack of dynamics, and to a certain degree, scale.

More recently, annotations embedded in HTML pages have become another prevalent source of structured data on the Web, building on standards such as RDFa[1], Microdata[2] and Microformats[3]. Markup is used by search engine

---

[1] RDFa W3C recommendation: http://www.w3.org/TR/xhtml-rdfa-primer/.
[2] http://www.w3.org/TR/microdata.
[3] http://microformats.org.

© Springer International Publishing AG 2016
A. González-Beltrán et al. (Eds.): SAVE-SD 2016, LNCS 9792, pp. 90–100, 2016.
DOI: 10.1007/978-3-319-53637-8_10

providers to interpret content of Web pages or enrich result pages with factual entity descriptions. One central effort is the schema.org initiative[4], driven by Google, Yahoo!, Bing and Yandex, aiming at defining a common vocabulary for describing entities on the Web and driving its adoption. A recent initiative [3] investigating a large-scale Web crawl from 2014 of 2.01 billion HTML pages constituting more than 15 million pay-level-domains (PLDs) found that 26% of all pages contain some form of embedded markup already, resulting in a corpus of 20.48 billion RDF quads[5].

Considering the apparent upward trend of adoption [1] (the proportion of pages containing markup increased from 5.76% to 26% between 2010 and 2014) and the still comparably limited nature of the investigated Web crawl, the scale of the data suggests a significant potential for exploiting it for a wide range of tasks, such as entity retrieval, knowledge base population or entity summarization.

Despite a growing interest in such embedded semantics, a thorough understanding of its adoption for scholarly resource metadata is still lacking. In this paper, we present the first study of scholarly data extracted from embedded annotations, utilizing the Web Data Commons as the largest crawl of embedded markup so far. Our analysis investigates questions about the level of adoption of terms and types, the shape and characteristics of entity descriptions and the distribution of data across the Web, for example, in terms of Pay Level Domains (PLDs), Top Level Domains (TLDs) or data publishers. In the following section we discuss the research questions and the methodology, followed by the data analysis and results of our study in the subsequent sections.

2   Methodology

2.1   Research Questions

The main target of this work is to answer certain questions regarding the usage of markup on scholarly data through a quantitative analysis. The research questions addressed in the following sections are:

- **RQ1**: *What are frequently used types and terms for scholarly data?* The main aim is to shape a better understanding of the adoption of vocabulary terms to comprehend the knowledge embedded through markup statements.
- **RQ2**: *How are statements about bibliographic data distributed across the Web and who are the key providers of bibliographic markup?* With this research question, investigated in Sect. 4, we research the distribution of data across domains and the indicated publishing institutions. We also aim to get a better understanding of the topic distribution, i.e., whether or not a strong bias towards particular scientific disciplines can be observed.
- **RQ3**: *What frequent errors can be observed?* In this context, we look into schema violations; significant syntactic and semantic errors introduced by data providers (Sect. 5).

---

[4] http://schema.org.
[5] http://www.webdatacommons.org.

These questions are approached through a quantitative analysis using the dataset described in the following section.

2.2 Methodology and Dataset

For our investigation, we use the Web Data Commons (WDC) dataset, being the largest available corpus of markup, extracted from the Common Crawl. Of the crawled web pages, 30% contain structured data which covers 17% pay-level-domains (PLDs)[6]. In addition, 20.48 billion RDF quads have been extracted, a significant amount when compared to DBpedia (4.58 million entities[7]) and Freebase (2.4 billion facts[8]). For our work we considered all statements which describe entities (subjects) that are of type *s:ScholarlyArticle* or of any type but co-occurring on the same document with any *s:ScholarlyArticle* instance.

To extract this subset, we processed the entire WDC2014 dataset using a Hadoop cluster for processing and extracting the investigated subset Our extracted dataset contains 6,793,764 quads, 1,184,623 entities, 83 distinct classes, and 429 distinct predicates. Due to space constraints, in later sections we will refer to *s:ScholarlyArticle* as *s:SchoArt*.

In our study, we have focused on schema.org as it is the most widely used schema and concentrated on *s:SchoArt*, *s:Person* and *s:Organization* for our analysis. Although there is a wide variety of types used for bibliographic and scholarly information, *s:SchoArt* is the only type which explicitly refers to scholarly articles. While this restricts our study with respect to recall, we followed this approach to enable a high precision of the analysed data within the scope of our study, where the goal is to provide conclusive insights into scholarly works markup only.

In order to identify related metadata to scholarly articles, our target was to find additional statements which relate the extracted instances of *s:SchoArt*. Since links between markup entities are sparse, the assumption that a node representing an author or affiliation of a specific article would be linked by the respective article instance does not hold true in the majority of cases. For this reason, we also consider instances of *s:Person* and *s:Organization* which co-occur with scholarly articles assuming that these will provide information about publishers or authors of the corresponding article.

3   Adoption of Scholarly Types and Predicates

This section addresses RQ1; we present an overview of utilized types and predicates in our extracted dataset. The major types considered are scholarly article (*s:SchoArt*), person (*s:Person*), and organization (*s:Organization*). Out of 6,793,764 triples and 1,184,623 entities, 280,616 instances are of *s:SchoArt*, 847,417 instances are of *s:Person* and 3,798 instances are of *s:Organization*.

---

[6] http://webdatacommons.org/structureddata/2014-12/stats/stats.html.
[7] https://en.wikipedia.org/wiki/DBpedia.
[8] https://en.wikipedia.org/wiki/Freebase.

Table 1. *Top-10* predicates used for *s:SchoArt*

Predicates	Occurrence
s:author	913,884
s:genre	204,954
s:image	191,879
s:headline	134,742
s:description	121,168
s:datePublished	119,448
s:publisher	115,896
s:keywords	104,488
s:name	78,873
s:editor	78,781

Table 2. *Top-10* PLDs according to the number of entities.

PLD	Entities	Statements
springer.com	850,697	3,011,702
bmj.com	106,777	877,589
mdpi.com	85,276	322,569
diabetesjournals.org	80,911	250,804
mendeley.com	42,564	143,376
biodiversitylibrary.org	24,946	122,457
gradesaver.com	24,108	121,592
grupoescolar.com	16,838	104,701
eurecom.fr	8,820	40,349
econjwatch.org	6,817	32,434

Among the organizations 1 instance is tagged as *s:Educational* organizations and 32 instances are tagged as *s:school* which is a further subtype of educational organization. Note that all the types and their subtypes are found by explicitly looking at the predicates for that particular type or subtype. For example, we have only captured those instances as *s:SchoArt* where the predicates corresponding to the instances specify scholarly article.

In Table 1, we present the *top-10* predicates ranked according to their occurrence. We find that for the type *s:SchoArt*, the predicate *s:author* depicts the highest occurrence with a frequency of 913,884. We also computed the number of distinct predicates for each instance for every extracted type. Figure 1 shows the distribution of distinct predicates over all the instances of the extracted types (*s:SchoArt*, *s:Person* and *s:Organization*). The number of distinct predicates for *s:SchoArt* varies from 1 to 14, for *s:Person* it from 1 to 9, and for *s:Organization* from 1 to 4.

It can be observed from both the distribution and the top-k predicates table, particular predicates are used very frequently, while there is a long tail of predicates which are hardly used. This provides insights as to the kind of knowledge which can be extracted from embedded scholarly data, where popular metadata is described in a fairly complete manner, for instance, author names and publication titles, while more specific information, for instance, about genres or publishers are less frequently found.

4   Distribution Across Domains and Documents

This section addresses RQ2, investigating the origins of bibliographic data, by analyzing the distribution of bibliographic markup across Pay-Level-Domains (PLDs), Top-Level-Domains (TLDs) and documents. There are 213 distinct PLDs, 38 TLDs and 199,979 documents across the subset.

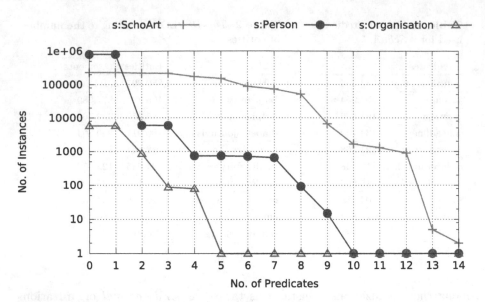

Fig. 1. Cumulative distribution of predicates over instances across extracted types. The number of instances (y-axis) are presented in log scale.

4.1   Distribution Across PLDs, TLDs and Documents

The distribution across domains and documents is represented in the plots of Fig. 2, where the blue (lower) line corresponds to the distribution of entities and the red (upper) line corresponds to the distribution of statements over PLDs, TLDs, and documents. The number of entities/statements presented on the *y-axis* are plotted in the logarithmic scale. As observed from the dataset, the number of statements is much higher than the number of entities corresponding to each PLD, TLD, or document. Another observation is the power law-like distribution of embedded markup across PLDs, TLDs, and documents, where

Table 3. Top-10 documents ranked according to embedded entities.

URL	Entities	Statements
<http://link.springer.com/article/10.1140%2Fepjc%2Fs10052-012-2183-y>	3843	7700
<http://link.springer.com/article/10.1007%2FJHEP02%282010%29041>	3035	6077
<http://www.russki-mat.net/page.php?l=FrFr\&a=C>	2486	9942
<http://link.springer.com/article/10.1140/epjc/s10052-010-1339-x>	2118	4242
<http://link.springer.com/article/10.1140/epjc/s10052-009-1227-4>	2114	4234
<http://link.springer.com/article/10.1140/epjc/s10052-010-1350-2>	2114	4234
<http://link.springer.com/article/10.1140%2Fepjc%2Fs10052-012-2175-y>	1999	4012
<http://www.chapman.edu/our-faculty/vernon-smith>	1879	5636
<http://cns.slis.indiana.edu/publications/>	1410	3507
<http://www.russki-mat.net/page.php?l=FrFr\&a=L>	1287	5144

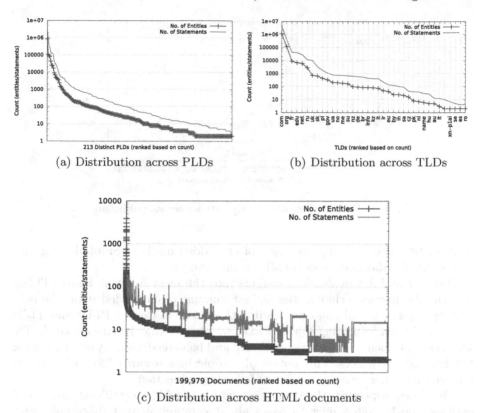

(a) Distribution across PLDs

(b) Distribution across TLDs

(c) Distribution across HTML documents

Fig. 2. Distribution of entities/statements over PLDs, TLDs and documents. (Color figure online)

usually a small amount of sources provide the majority of entities and statements.

In Fig. 2(a) we plot the different PLDs along *x-axis* and the number of entities/statements corresponding to each PLD along *y-axis* in the logarithmic scale. We represent the PLDs in the ranked order of the number of entities and statements corresponding to them. For example, springer.com exposes a total of 850,697 entities and 3,011,702 statements. A detailed list of the *top-10* PLDs is shown in Table 2.

In Fig. 2(b) we plot the different TLDs along the *x-axis* and the number of entities/statements corresponding to each TLD along the *y-axis* in the logarithmic scale. For example, documents from .*com* domains expose 1,139,436 entities and 4,640,718 statements. As can be observed, .*com* and .*net* URLs are very frequent, while some national TLDs such as .*fr* appear among the top providers of scholarly bibliographic data. Basing our study on the assumption that the Common Crawl is a representative Web crawl, this provides first insights into the early adopters of embedded scholarly markup. A deeper look into the *top-k* PLDs supports the assumption that French academic and library institutions

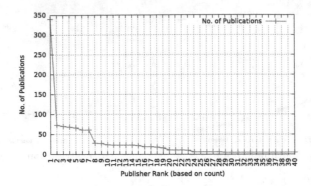

Fig. 3. Distribution of scholarly articles across publishers.

seem to be among the top providers of embedded markup. Similarly, Fig. 2(c) shows the distribution across HTML documents.

Tables 2 and 3 provide some insights into the most frequent PLDs (TLDs) and the documents including the highest amount of embedded data. We note that springer.com and *.com* are leading the queue in case of PLDs and TLDs respectively. On inspecting *top-10* PLDs, we observe that journals from the life sciences field, such as diabetesjournals.org and biodiversitylibrary.org are among the key data providers. This notion of a topic bias towards life and medicals sciences is further investigated in the following subsection.

On closer inspection, the documents which provide a significant amount of entities (top-10) often refer to pages about comprehensive publications, such as a book publication with rich annotation of bibliographic data, such as references for each chapter, as in the case of <http://link.springer.com/article/10. 1140%2Fepjc%2Fs10052-012-2183-y> with 3843 embedded entities. Note that in rare cases (for instance, <http://www.russki-mat.net/page.php?l=FrFr&a=C> in row 3, referring to a Russian slang dictionary), flawed data is included, where instances are incorrectly typed and are actually not referring to scholarly data. This calls for further investigation into the correctness of embedded data (also see Sect. 5).

4.2 Distribution Across Topics and Publication Types

In order to better understand the topic coverage of scholarly data, we provide some initial insights into the most frequent publishers of detected scholarly articles, as indicated by the data itself, and the suspected topic bias of articles themselves. In Fig. 3 and Table 4 we show the overall distribution of scholarly articles across different publishers (533 distinct publishers in total) and the top-10 publishers ranked according to their publication count.

Similar to the distribution across PLDs, TLDs, and documents, the spread across publishers follows a power law distribution.

As observed in the table, most publishers seem to be either from the Computer Science domain (IEEE, Telecom Paris) or seem to be cross-domain, with a

Table 4. Top-10 publishers and their publication counts.

SchoArt:Publisher	#Publication
Econ Journal Watch	340
IEEE@fr	73
IEEE@en	70
TELECOM ParisTech@fr	68
TELECOM ParisTech@en	66
ENST Paris@en	61
ENST Paris@fr	61
Universit de Nice@fr	28
Universit de Nice@en	27
Springer@fr	24

Table 5. Most frequent publication types across the WDC dataset

SchoArt:genre	Article count
Article@en	7,788
Thesis@fr+@en	373
Conference@en+@fr	188
Journal@fr+@en	115
Rapport@fr+@en	16
Ouvrage@fr	7
Poster@en+@fr	8
Book@en	5
Talk@fr+@en	6
HDR@en+@fr	2
Others	6

Table 6. Top-10 article titles (pre-cleaned) ranked according to their occurrence.

Article title (SchoArt:name)	Occurrence
Highlights From the Latest in Diabetes Research@en	39
Essential information about patterns of victimisation among children with disabilities@en	36
Whose Oxis Being Gored? When Attitudinalism Meets Federalism@en	36
People with unhealthy lifestyle behaviours benefit from remote coaching via mobile technology@en	27
Longer duration of exclusive breastfeeding associated with reduced risk of childhood asthma up to age six@en	25
People with diabetes and selfreported severe hypoglycaemia have increased mortality risk over years@en	25
Community based nonpharmacological interventions delivered by family caregivers reduce behavioural and psychological symptoms of dementia@en	24
Preoperative physical therapy reduces risk of postoperative at elect as is and pneumonia in people undergoing elective cardiac surgery@en	24
How to Choose the Least Unconstitutional Option:Lessons for the President(and Others)from the Debt Ceiling Standoff@en	24
Post menopausal women with medically treated diabetes have increased risk of lung cancer@en	22

particular bias towards Life Sciences related literature (e.g. Springer). In order to get a clearer understanding of the actual topics of articles, we inspected the titles (*s:name*, *s:headline*) of articles. Although titles are often not well-populated we investigated frequently occurring titles, and ignored obviously noisy or misleading annotations.

From Table 6 we note that the top-10 actual article titles are all from the biological or medical domain, further indicating a strong inclination towards the Life Sciences.

In addition, we investigated the genre (*s:genre*) of detected articles, meant to describe the publication type. In Table 5, we cluster the genres such as thesis and journals having *@en* and *@fr* tags together to enhance readability. While articles (*Article(@en)*) seem by far the most used genre annotations, the whole range of bibliographic types is covered. Observed language annotations again confirm some bias towards English and French content and data providers.

5   Frequent Errors: Schema Violations

Errors are frequently found in embedded annotations, and the extent varies depending on the type of error. For instance, the use of undefined types and predicates is more frequent in traditional Linked Data, due to the fact that errors propagate through a dataset, as opposed to embedded data [4]. Other error types, such as schema violations and misuse of object properties are particularly frequent in embedded data. In Table 7, we report the most frequently misused predicates, that is predicates which are defined as object property but refer to data type/literal or vice versa. Here S and P are used as to indicate the range of the property, either <http://schema.org/ScholarlyArticle/> or <http:// schema.org/Person/> respectively. For example *S:author* is an object property

Table 7. Top 10 misused predicates. Range refers to the expected range according to the schema.org vocabulary definition and is either *OP*-Object Property or *DP*-Data Type Property

Predicates	Occurrence	Range	Object	Data type	%Error
S:publisher	144147	OP	997	143150	99.31
S:creator	44615	OP	28550	16065	36.01
S:author	1048110	OP	697024	351086	33.49
S:about	888	DP	97	791	10.92
P:dateModified	7644	DP	419	7225	5.48
S:sourceOrganization	1637	OP	17	1620	1.01
P:affiliation	2144	OP	2129	15	0.69
S:headline	145953	DP	413	145540	0.28
S:datePublished	127494	DP	76	127418	0.06
P:editor	78781	OP	78773	8	0.01

having 1,048,110 occurrences within the dataset, where 697,024 instances correctly refer to a node (object), while the remaining 351,086 instances use it as a datatype property, directly referring to a literal (error rate 33.49%). From the Table 7 we can also observe that most often object properties are violated, while data type properties are largely compliant. This observation, further highlighting the lack of explicit links (object references) between entity descriptions in embedded markup, suggests that further research into coreference resolution and entity interlinking is required, in order to utilize scholarly markups as a potential knowledge graph.

6 Conclusion and Future Work

In this work, we have provided a first study on the coverage and characteristics of bibliographic metadata embedded in Web pages. Insights are provided with respect to frequent data providers, the adoption and usage of terms and the distribution across providers, domains and topics. The distribution in all cases follows a power law, with few providers and documents contributing the majority of data. The same can be identified for vocabulary terms, where few predicates are highly used, complemented by a long tail of predicates which are only used to a very small extent. With regard to the distribution across domains, a certain bias towards French data providers is observed based on manual investigation of the top-k genres and publishers. Article titles, PLDs, and publishers suggest a bias towards specific disciplines, namely Computer Science and the Life Sciences. However, the question as to what extent this is due to the selective content of the Common Crawl or representative for schema.org annotations on the Web in general, requires additional investigation.

As a part of future work, we are planning to conduct a follow-up study using a targeted crawl of typical providers of scholarly data (publishers, academic organizations, libraries), which would enable a more exhaustive and representative analysis. By limiting ourselves to explicitly annotated scholarly articles, it is also worth highlighting that a significant amount of bibliographic data has been excluded from our study. Here, as part of future work, other methods should be taken into account to classify implicitly typed bibliographic or creative work into scholarly or non-scholarly works. In addition, resolution of co-references and research into specifically tailored entity interlinking mechanisms would help to provide a more consolidated picture of the scholarly knowledge graphs which can be extracted from embedded data. This is an area where we see some key opportunities for related future work. Extracting (scholarly) knowledge graphs from Web documents provides opportunities for generating data far beyond the scale and dynamics of traditional datasets in the area. At the same time, embedded (scholarly) data can provide invaluable training data for targeted, i.e. domain-specific information extraction and linking algorithms for scholarly information.

References

1. Bizer, C., Eckert, K., Meusel, R., Mühleisen, H., Schuhmacher, M., Völker, J.: Deployment of RDFa, microdata, and microformats on the web – a quantitative analysis. In: Alani, H., et al. (eds.) ISWC 2013. LNCS, vol. 8219, pp. 17–32. Springer, Heidelberg (2013). doi:10.1007/978-3-642-41338-4_2
2. Dietze, S., Taibi, D., dAquin, M.: Facilitating scientometrics in learning analytics and educational data mining the LAK dataset. Seman. Web J. (2015)
3. Meusel, R., Petrovski, P., Bizer, C.: The webdatacommons microdata, RDFa and microformat dataset series. In: Mika, P., et al. (eds.) ISWC 2014. LNCS, vol. 8796, pp. 277–292. Springer, Cham (2014). doi:10.1007/978-3-319-11964-9_18
4. Meusel, R., Paulheim, H.: Heuristics for fixing common errors in deployed *schema.org* microdata. In: Gandon, F., Sabou, M., Sack, H., d'Amato, C., Cudré-Mauroux, P., Zimmermann, A. (eds.) ESWC 2015. LNCS, vol. 9088, pp. 152–168. Springer, Cham (2015). doi:10.1007/978-3-319-18818-8_10

Semantic Technologies for Citation and Topic Analysis

Citation Functions for Knowledge Export - A Question of Relevance, or, Can CiTO Do the Trick?

Joakim Philipson[✉]

Kungliga Biblioteket - National Library of Sweden,
Information Systems Department, Stockholm, Sweden
joakim.philipson@kb.se

Abstract. This paper explores the possibility of promoting knowledge export by means of citation function indexing using *CiTO*, the Citation Typing Ontology [4]. Instances of knowledge export are exemplified by cross-disciplinary citations, which, it is suggested, may indicate a prolonged life time use of documents. For CiTO to serve the purpose of promoting knowledge export, it should be more specific about citation functions, separating them from evaluation, and then be put to test as a discovery tool.

Keywords: Citation functions · Cross-disciplinarity · Knowledge export · Relevance

1 Introduction: The Relevance of Citation Functions

Citations can often be seen as observable results of a transfer of knowledge, as records of used information. Citations as a potential measure of relevance was noted at least implicitly by *Gilbert* [13]:116. However, the use of citations vary greatly. We focus here in particular on cross-disciplinary citations and the different functions they fulfil. What purpose do they serve? We want to know how the cited information is used in the citing context, fully aware that there may be other reasons behind citations than strictly intra-scientific judgments of relevance, e.g. as a purely rhetorical device [13,21]. The references ultimately appearing in an article may also be determined by factors outside the author's immediate control, such as peer review and journal policies.

Still, why is it that certain documents are being found relevant for the most various purposes over and over again long time after their publication, while others tend to fall into oblivion only a few years after their appearance? Which factors are involved in distinguishing the potentially long-lived cited document from the less successful, more short-lived ones? *Van Raan* and more recently *Ke et al.* studied so called "sleeping beauties" in science, i.e. instances of "a publication that goes unnoticed (sleeps) for a long time and then, almost suddenly, attracts a lot of attention (is awakened by a prince)." [26] Studying long

© Springer International Publishing AG 2016
A. González-Beltrán et al. (Eds.): SAVE-SD 2016, LNCS 9792, pp. 103–112, 2016.
DOI: 10.1007/978-3-319-53637-8_11

'sleeping beauties' (SBs) for the purpose of identifying cross-disciplinary citation functions promises to be rewarding, since "top SBs achieve delayed exceptional importance in disciplines different from those where they were originally published." [16] *Levitt and Thelwall* [17] found a link between multi-disciplinarity and a high citedness rate. However, their study did not address the question of cross-disciplinary knowledge export. Multi-disciplinarity and even more so interdisciplinarity or transdisciplinarity have more to do with the *integration* or *synthesis* of scientific disciplines working on a common research project, as in the emerging so called I2S, Integration and Implementation Sciences [8]:322. Cross-disciplinarity, on the other hand, is more about researchers in one scientific discipline seeking to apply new methodologies, solutions or problems taken from another, sometimes very distant discipline. Thus, results from studies of interdisciplinarity, transdisciplinarity or multi-disciplinarity cannot automatically be applied to cases of cross-disciplinary knowledge export. By knowledge export we understand here the transfer of knowledge from one discipline to another as documented by cross-disciplinary citations.

Apart from the phenomenon of sleeping beauties, citation analyses have shown substantial variations in citation patterns over time from one discipline to another. There are indications e.g. that documents within the social sciences continue to be cited for a longer period of time than what is the case for the natural sciences [12]. However, there are also examples of remarkably long-lived documents from the natural sciences. A classic paper by *Albert Einstein* from 1906 was still being cited in journal articles within fields so diverse as dairy sciences, pharmacology, physiology, ceramics, water pollution, acoustics, fluid mechanics, sedimentary petrology and molecular biology during the 1960s [12], and well into this century again within ceramics, mechanics and sedimentology.

Another example is that of *Molina and Rowland* [22], a paper from the field of atmospheric chemistry published in 1974, which has continually been cited at least up until the mid 1990s also within disciplines such as computer science, law, management, ophthalmology, optics, political science, pharmacology, sociology, and, even more recently, risk management and medicine. Noteworthy in cases like these, where papers continue to be used and cited over a long period of time, is precisely the subject dispersion of citing papers. In the case of [22], the fact that the paper was published in a prestigious multi-disciplinary scientific journal like *Nature* most likely promoted its exposure also to scientists from outside atmospheric chemistry. The attention it received was no doubt renewed in 1995 when the authors, together with Paul Crutzen, were awarded the Nobel prize "for their work in atmospheric chemistry, particularly concerning the formation and decomposition of ozone".

Still, most articles published in *Nature* never come near the very high citation score attained by this paper. Moreover, [22] received most citations years after its publication, not while it was still new and "outsiders", with a fresh issue of Nature in hand, were more likely to be "accidentally" exposed to the paper, but still before the Nobel prize award (although admittedly there was a new peak in its citation count in 1995, still lower though than in the top year 1976).

Understanding the multipurposeness of scientific papers and their potential for knowledge export calls for an explanation of the function that the cited source fulfils in the context of the citing documents. How does the cited information fit into this sometimes completely new disciplinary environment? In this paper we examine a few examples of cross-disciplinary citation functions, to see if they could also be expressed by the emerging standard citation typing ontology *CiTO* [4] for the purpose of promoting knowledge export.

2   Content-Based Citation Analysis and Citation Functions

Most citation analysis studies so far have been quantitative. Citation counts have been made, e.g., in order to identify the core literature of a scientific discipline and co-citation clustering has been used for mapping the structure of scientific disciplines [12]. *Lipetz*, pioneer of *qualitative* citation analysis, investigated the relationship between cited reference and citing document, aiming to improve the selectivity of citation indexes, but the 29 categories he proposed were obviously not intended to constitute a final judgment on the matter [18].

Since then qualitative or *content-based citation analysis* [11] studies have produced a multitude of different schemes describing the various functions of citations, with considerable overlap between categories, although the exact labels used for classification differ among authors [19].

The earlier classification schemes for citation functions relied essentially on manual citation analysis of relatively small sets of articles (typically 10 to 100 items), while later attempts have been made to use semi-automated or computational methods for citation classification of larger samples of full-text articles. An overview of these attempts is found in *Ding et al.* [11].

However, automated methods for citation classification, relying on explicit signals or cue words for identification of citation functions [6], may not capture more complex cross-disciplinary citation relationships of the syntagmatic kind described by *Green and Bean* [14], where the relevance of the cited source to the citing document stems rather from the provision of a missing piece of information serving e.g. as part of an evidence chain. An example of this kind of relationship is given in the next section where we will be looking closer at some cross-disciplinary citations apparently representing instances of knowledge export. Thus, this paper still depends on a small number of manually extracted citations from a limited set of articles. The purpose is simply to understand why a scientific article was found useful also outside its original field of research. What follows are some selected examples of citations of papers from the field of atmospheric chemistry or stratospheric ozone monitoring [24], all introduced by a description of the identified citation function followed by an analysis and discussion of a possible application of CiTO object properties.

3   Cross-Disciplinary Citation Functions

Comparison: Citation refers to similar results from another field of research. It may appear as a metaphorical type of relation, in which "one complex unit is perceived as being structurally equivalent (as a whole or in part) to another" [14]:660. The importance of analogical, structural comparison (of similar or dissimilar elements) for knowledge transfer has been extensively described by *Day and Goldstone* [10]. So it seems only natural that it figures in cases of cross-disciplinary knowledge export and use of scientific data another field of research. A possible instance of this type is [2]:

> "Similar long-term trends are to be found in total column ozone measurements.... London and Kelley (1974) examining global total ozone found an increase in both the Northern and the Southern Hemisphere during the 1960s."

This article had at the time of access no shared subject descriptors with the cited document [20] in two different research databases (Aerospace database[1], accessed March, 1996, and the Pascal database[2], using exclusively English descriptors.) Thus, this is not a case of *topic matching*. However, the citation link between [2,20] appears to be rather strong, with the citation providing both *measurement data*, functioning as an item of *comparison* and lending supporting *evidence* together with other cited documents to the conclusion that

> "the long-term trend in stratospheric water and its similarity to the long-term trend in stratospheric ozone suggest that these changes arise from long-term changes in the intensity of the circulation." [2]:2164

But obviously, the article is not *about* stratospheric ozone variation, which is the topic of [20]. The main topic of [2] is described by the title: *Stratospheric water vapor variability for Washington, DC/ Boulder, CO: 1964-82*. Citations for this type of "non-topical" comparisons seem difficult to represent by means of CiTO. A possible candidate for a suitable CiTO object property in this case would perhaps be *cito:extends*[3] [20], but it does not capture accurately the "non-topical" quality of this instance.

Evidence: Citation is used for support of propositions in citing entity. Instances of conclusive, logically binding proofs may be rare; rather, reference is often to the apparent agreement between measurement data and predictions of a theory or a model. This type of citation might seem more natural for specialists within a narrower field of research, as it may sometimes require expertise in the field to seize the arguments involved. However, there are also clear examples of cross-disciplinary citations for evidence. Consider the following extract from an article published in a botanical journal as an illustration:

---

[1] http://media2.proquest.com/documents/pq_advanced_tech_aerospace_prof_prosheet.pdf.

[2] http://media2.proquest.com/documents/pascal.pdf.

[3] http://purl.org/spar/cito#, http://purl.org/spar/cito/extends.

"Good estimates of the present stratospheric distribution of ozone and subsequent UV radiation are known (Koller 1952; Dtsch 1969; Cutchis 1974). The total amount of ozone in the northern hemisphere is maximal in spring and minimal in fall. ... It is suggested that among flowering plants of the northern hemisphere, many of which have white or yellow flowers (Table 2), there has been convergent evolution in floral UV absorption. Yellow and white flowers are high in flavonoid pigments which strongly absorb UV light. The seasonality of UV radiation may be one major selective pressure. Yellow and white flowers comprise as much as 85% of an arctic flora (Kevan 1972)." [25]:26f

Discerning some of the more important of premisses involved in the inference leading to the hypothesis in the third sentence of the extract, there is first the observation of the seasonal variation of stratospheric ozone and the subsequent seasonal variation of ultraviolet radiation reaching the earth, leading to a spring maximum of stratospheric ozone and a subsequent spring minimum of UV-radiation in the northern hemisphere (since stratospheric ozone absorbs UV-radiation). Then there is the knowledge that yellow and white flowers are strong absorbants of UV-radiation. Finally there is the evidence of the predominance of yellow and white flowers in the northern hemisphere. Together these premisses make probable the hypothesis that UV-absorption ability has acted as a selective evolutionary mechanism for flowers in the northern hemisphere. It is important to note here that the different premisses come from different subject areas. The first three cited sources in the extract belong to geophysics or climatology, whereas *(Table 2)* and *(Kevan 1972)* are from botany. Despite the differences in subject, the premisses apparently "fit" together, as "slots in a framework" [14]:660. One describes certain environmental conditions. Another describes an important property of the object being studied, influencing its adaptation to the conditions described by the first. The third premiss describes the frequency of occurrence of the object being studied, thereby corroborating the importance of the property described by the second premiss. Together they make up an evidential structure, that accounts for the relevance of the cited entities to the purpose of the citing document. Thus, all the cited entities here could apparently be ascribed the CiTO object property *cito:isCitedAsEvidenceBy*[4] [25]. Alternatively, some of these citations, e.g. those of the strictly botanical sources, might also be described by the CiTO property *cito:isCitedAsDataSourceBy*[5] [25].

Force: Citation refers to a likely *structure*, *mechanism* or *cause* behind observed phenomena. A typical example is a reference to a chemical reaction described by the cited entity. Again this type of citation function would seem to be essentially an internal affair among specialists within a field of research, but examples of "outsiders" making use of it also occur, as this excerpt from a medical journal illustrates:

---

[4] http://purl.org/spar/cito#, http://purl.org/spar/cito/isCitedAsEvidenceBy.

[5] http://purl.org/spar/cito#, http://purl.org/spar/cito/isCitedAsDataSourceBy.

"Stratospheric ozone depletion, accompanied by increases in ambient, bio-
logically destructive ultraviolet-B radiation, [104]may exacerbate the effect
of climate change on infectious diseases. Arising from a different anthro-
pogenic process than climate change, ozone destruction is occurring primarily from reactions between ozone and halogen free radicals derived from
chlorofluorocarbons, other halocarbons, and methyl bromide."[105] [23]; ref.
(105) is to [22]

No specific object property was found in CiTO for citations referring to a
likely cause, mechanism or explanatory force. A significant difference between
the *evidence* and the *force* citation functions appeared in [24], where the 32
citations of [22] for *evidence* had a median publishing year of 1975, only one year
after the cited source, whereas the 26 citations of the *force* type appeared to be
among the most "long-lived", in the sample, with a median publishing year of
1984, ten years after the cited source. The sample in that study was too small
to allow any definite conclusions, but the apparent difference in age distribution
may not be surprising anyway. The reference to an explanatory force in the form
of a chemical reaction or structure should be of such permanence that it can
be expected to be found not only in articles in scientific journals, but even in
textbooks.

Method: Citation refers to the method employed in the cited work. This does
not necessarily mean that the same method is used or even advocated by the
citing article, as observed in the following example:

"Total ozone data were previously analyzed by a number of authors including Angell and Korshover (1973), London and Kelly [sic!] (1974) with particular interest in quantifying long-term trends. The statistical procedure
commonly used in these studies is linear regression analysis (i.e. fitting a
straight line) applied to adjusted total ozone values (e.g. deviations from
monthly means ...). However, problems arise in the interpretation of results
from these linear regression models since these models fail to take account
of the positive autocorrelation that is present in the ozone data. Hence, we
consider time series analysis that accounts for autocorrelation in a quantitative trend assessment of ozone data." [7]:460)

In CiTO, the object property relating to method presupposes that the cited
method is actually used by the citing document, *cito:usesMethodIn*[6]. This is a
problematic feature of CiTO; while some properties seem to be too general to
distinguish between different specific citation functions, other properties, like
this one, presuppose an active use or endorsement of the *content* of the citation
function extracted from the cited entity. There are of course a number of other
object properties in CiTO expressing a negative evaluation of the cited entity, but
these are again more general and hold no information about which function or
part of the cited entity that is negatively evaluated. The methodological citations

---

[6] http://purl.org/spar/cito#, http://purl.org/spar/cito/usesMethodIn.

in the aforementioned study [24] were few in number, but their relatively "long life" might be more than just an accidental effect of the selection. If so, support could be gained from the results of [5], showing how a scientific paper that was formerly frequently cited for "theoretical" reasons as describing the structure of collagen suddenly ceased to be among the highly cited papers for a short time, when the focus of research in the field shifted from structural studies to biosynthesis, only to reappear as one of the high ranking cited sources a year later, but then cited rather for its methodology [12]:127f.

Result: Citation involves an implication, viz. "*if* information C_0 contained in cited document is true, and if furthermore conditions C_1, C_2, ... C_n hold good, *then* the consequences will be such and such". Hence, the citing article does not necessarily have to endorse a claim of truth for the cited information; the only claim is for the potential *result*, given the conditions described by the antecedent of the implication. The auxiliary conditions C_1, C_2, ... C_n furthermore do not have to be topically related to the cited information. The only requirement is that there must be no contradiction among them. In [24] several instances of this type of citation appeared in articles from journals, that were clearly peripheral to the field of research concerned with stratospheric ozone monitoring, coming from such disciplines as molecular biology, botany, or ophthalmology. Researchers from "outside" naturally should be more concerned with the implications of the cited information for their own field of research, rather than with trying to assess the validity of that information, lacking the necessary specialist competence for that. The following passage may serve as an example:

"Recent studies by Cicerone (4) and *Molina and Rowland* (7) state that increased use of fluorocarbons in aerosols and refrigerants could severely deplete the protective layer of ozone in the stratosphere. This would increase the level of UV-B radiation reaching the earth's surface. ... The object of this study was to determine the effects of UV-B irradiation on local lesion development of Chenopodium quinoa Willd. 'Valdivia' plants inoculated with potato virus S (PVS)." (*Semeniuk and Goth* [3]; ref. (7) is to [22])

Cito has an object property *cito:usesConclusionsFrom*[7] that might fit for this kind of citation function, but again it seems the CiTO object property presupposes an active claim of truth for the cited information, whereas the **result** function described here is more neutral and conditional. In general it would be preferable to separate citation functions from evaluative judgement as clearly as possible, so that each citation function identified could be given one of three values, positive (+), negative (−) or neutral (0).

Now, as we have seen, not all the above examples of citation functions are directly translatable into CiTO object properties, but they nevertheless shed some light on the use of scientific information outside the discipline whence it originated. Possibly other, even more compelling examples such as these can be

---

[7] http://purl.org/spar/cito#, http://purl.org/spar/cito/usesConclusionsFrom.

found, where the age distance between cited and citing documents is larger, as we already saw in Sect. 1 for *Einstein (1906)* and [22].

4   Conclusions: Indexing for Knowledge Export - Can CiTO Do the Trick?

Could citation indexing with *CiTO* [4] serve the purpose of "knowledge export"? From the examples above it appears CiTO is not specific enough to capture the finer differences between citation functions. At the same time there seems to be some redundancy in the present version of CiTO [9], so having index terms more accurately describe citation functions while separating them from value judgments, does not necessarily imply that the number of object properties would have to grow substantially.

We have seen some instances of cross-disciplinary citations characterized by the kind of *hierarchical* or *structural, syntagmatic* relationships between citing and cited source, described by *Green and Bean* [14]. With the citing entity representing the user need, "the topic of the user need and the topic of the cited passage are related as class and subclass, or... as class and class-member" [14]:659. This kind of "type-token" relationship can be expressed in citations by the provision of an instance of the class referred to. It may also appear in the form of the citation function referred to above as *comparison* with a structurally equivalent unit.

Structural (or *syntagmatic*) relationships are those where "the topic of the cited passage corresponds to a component within a conceptual syntagmatic structure (...), while the topic of the user need corresponds to another component within the structure, or again, the structure at large" [14]:660. We saw an example of this relationship in the *evidence* function in the case of [25] above.

The limited importance of topic matching relationships in citations was confirmed in a study by *Harter et al.* [15] from the area of library and information science, in which the subject similarity among pairs of cited and citing documents was found to be very small. However, independence from topic matching may vary between disciplines. *Guerrero-Bote et al.* found a significant correlation between the knowledge export and import rates of different subject categories: "This indicates that there are Subject Categories which are more independent, importing and exporting little knowledge, and others with greater flows of knowledge across subject boundaries." [1]:440

Indexing citation functions is not so much about representing "mental models" or capturing the original "intention" of the citing author [6,9], but rather about describing the actual and potential *use* - past, present and future - of document contents. It is essential then to look at both sides of citation relationship simultaneously, the *citing* entity and the *cited* source. A combination of citation functions and subject headings, extracted from both citing and cited entities might offer even better prospects for knowledge export and provide researchers and readers with new context, adding new relevance to old documents, opening new opportunities for "evidence mining". What is needed is a proper test of the capability of an indexing system of citation functions like CiTO, possibly revised and revamped, to serve as a *discovery tool* across scientific disciplines. Preparation for such a test could perhaps start by indexing a sample of outside 'princes', who have awakened some of those long 'sleeping beauties', and then have a panel of independent researchers, unknowing of her history, find their way to "la Belle au bois dormant".

The resulting indexing scheme of a conclusive test should be sufficiently easy to use, so that virtually anyone who reads and writes and cites would be able to contribute to the indexing effort. Online publishers of scientific journals, managers of digital repositories like JSTOR and existing citation indexes like the Web of Science and CiteSeerX could make it happen by means of crowd-sourcing from the users. Ideally, tagging a scientific article online with citation functions from a controlled index language should be just little more complicated than liking a post on social media.

References

1. Guerrero-Bote, V., Zapico-Alonso, F., Espinosa-Calvo, M., Gómez-Crisóstomo, R., Moya-Anegón, F.: Import-export of knowledge between scientific subject categories: the iceberg hypothesis. Scientometrics **71**(3), 423–441. http://dx.doi.org/10.1007/s11192-007-1682-3
2. Mastenbrook, H.J., Oltmans, S.J.: Stratospheric water vapor variability for Washington, DCIBoulder, CO: 1964-82. J. Atmos. Sci. **40**(9), 2157–2165. http://dx.doi.org/10.1175/1520-0469(1983)040⟨2157:SWVVFW⟩2.0.CO;2
3. Semeniuk, P., Goth, R.W.: Effects of ultraviolet irradiation on local lesion development of potato virus S on Chenopodium Quinoa 'Valdivia' leaves. Environ. Exp. Bot. **20**(1), 95–98 (1980). http://dx.doi.org/10.1016/0098-8472(80)90224-5
4. Shotton, D., Peroni, S., Ciccarese, P., Clark, T.: CiTO, the Citation Typing Ontology (2015). http://purl.org/spar/cito/
5. Small, H.G.: A co-citation model of a scientific specialty: a longitudinal study of collagen research. Soc. Stud. Sci. **7**, 139–166 (1977). http://www.jstor.org/stable/284873
6. Teufel, S., Siddharthan, A., Tidhar, D.: Automatic classification of citation function. In: Proceedings of the 2006 Conference on Empirical Methods in Natural Language Processing (EMNLP 2006), pp. 103–110 (2006). http://www.aclweb.org/anthology/W/W06/W06-1613.pdf
7. Tiao, G.C.: Use of statistical methods in the analysis of environmetal data. Am. Stat. **37**(4b), 459–470 (1983). http://dx.doi.org/10.1080/00031305.1983.10483166
8. Bammer, G.: Change! Combining Analytic Approaches with Street Wisdom. Australian National University. ISBN: 9781925022650(ebook)
9. Ciancarini, P., Di Iorio, A., Nuzzolese, A., Peroni, S., Vitali, F.: Cognitive issues. http://dx.doi.org/10.1007/978-3-319-07443-6_39
10. Day, S.B., Goldstone, R.L.: The import of knowledge export: connecting findings and theories of transfer of learning. Educ. Psychol. **47**(3), 153–176 (2012). http://dx.doi.org/10.1080/00461520.2012.696438
11. Ding, Y., Zhang, G., Chambers, T., Song, M., Wang, X., Zhai, C.: Content-based citation analysis: the next generation of citation analysis. J. Assoc. Inform. Sci. Technol. **65**(9), 1820–1833. http://dx.doi.org/10.1002/asi.23256

12. Garfield, E.: Citation indexing: its theory and application in science, technology, and humanities. ISBN: 0-471-02559-3

13. Gilbert, G.N.: Referencing as persuasion. Soc. Stud. Sci. **7**, 113–122 (1977). http://dx.doi.org/10.1177/030631277700700112

14. Green, R., Bean, C.A.: Topical relevance relationships: I why topic matching fails; II. an explanatory study and preliminary typology. J. Am. Soc. Inform. Sci. **46**(9), 646–662. http://dx.doi.org/10.1002/(SICI)1097-4571(199510)46:9⟨646::AID-ASI2⟩3.0.CO;2-1, http://dx.doi.org/10.1002/(SICI)1097-4571(199510)46:9⟨654::AID-ASI3⟩3.0.CO;2-3

15. Harter, S.P., Nisonger, T.E., Weng, A.: Semantic relationships between cited, citing articles in library, information science journals. J. Am. Soc. Inform. Sci. **44**(9), 543–552 (1993). http://dx.doi.org/10.1002/(SICI)1097-4571(199310)44:9⟨543::AID-ASI4⟩3.0.CO;2-F

16. Ke, Q., Ferrara, E., Raddichi, F., Flammini, A.: Defining and identifying Sleeping Beauties in science. PNAS **112**(24), 7426–7431. http://dx.doi.org/10.1073/pnas.1424329112

17. Levitt, J.M., Thelwall, M.: The most highly cited Library and Information Science articles: interdisciplinarity, first authors and citation patterns. Scientometrics **78**(1), 45–67 (2009). http://dx.doi.org/dx.doi.org/10.1007/s11192-007-1927-1

18. Lipetz, B.-A.: Improvement of the selectivity of citation indexes to science literature through inclusion of citation relationship indicators. Am. Documentation **16**(2), 81–90 (1965)

19. Liu, M.: Progress in documentation - the complexities of citation practice: a review of citation studies. J. Documentation **49**(4), 370–408. http://dx.doi.org/10.1108/eb026920

20. London, J., Kelley, J.: Global trends in total atmospheric ozone. Science **184**(4140), 987–989 (1974). http://dx.doi.org/10.1126/science.184.4140.987

21. Moed, H.F.: Citation Analysis in Research Evaluation. Springer, Dordrecht (2005). ISBN 9781402037146 (ebook). http://dx.doi.org/10.1007/1-4020-3714-7

22. Molina, M.J., Rowland, F.S.: Stratospheric sink for chlorofluoromethanes: chlorine atom catalysed destruction of ozone. Nature **249**(5460), 810–812. http://dx.doi.org/10.1038/249810a0

23. Patz, J.A., Epstein, P.R., Burke, T.A., Balbus J.M.: Global climate change and emerging infectious diseases. JAMA **275**(3), 217–223 (1996). http://dx.doi.org/10.1001/jama.1996.03530270057032

24. Philipson, J.: The relevance of citations: a case study of stratospheric ozone monitoring. ISSN: 1401–5358 (1996). http://hdl.handle.net/2320/13707

25. Utech, F.H., Kawano, S.: Spectral polymorphisms in angiosperm flowers determined by differential ultraviolet reflectance. The botanical magazine = Shokubutsu-gaku-zasshi **88**(1), 9–30 (1975). http://dx.doi.org/10.1007/BF02498877

26. Van Raan, A.F.J.: Sleeping beauties in science. Scientometrics **59**(3), 467–472 (2004). http://dx.doi.org/10.1023/B:SCIE.0000018543.82441.f1

Semantic User Profiles: Learning Scholars' Competences by Analyzing Their Publications

Bahar Sateli[1](✉), Felicitas Löffler[2](✉), Birgitta König-Ries[2](✉),
and René Witte[1](✉)

[1] Semantic Software Lab, Department of Computer Science and Software
Engineering, Concordia University, Montréal, Canada
{sateli,witte}@semanticsoftware.info
[2] Department of Mathematics and Computer Science,
Friedrich Schiller University Jena, Jena, Germany
{felicitas.loeffler,birgitta.koenig-ries}@uni-jena.de

Abstract. Semantic publishing generally targets the enhancement of
scientific artifacts, such as articles and datasets, with semantic metadata. However, smarter scholarly applications also require a better model
of their *users*, in order to understand their interests, tasks, and competences. These are generally captured in so-called *user profiles*. We investigate a number of existing linked open data (LOD) vocabularies and propose a description of scientists' competences in LOD format. To avoid the
cold start problem, we suggest to automatically populate these profiles
based on the publications (co-)authored by users, which we hypothesize
reflect their research competences. Towards this end, we developed the
first complete, automated workflow for generating semantic user profiles
by analyzing full-text research articles through natural language processing. We evaluated our system with a user study on ten researchers from
two different groups, resulting in mean average precision (MAP) of up to
92%. We also analyze the impact of semantic zoning of research articles
on the accuracy of the resulting profiles. Finally, we demonstrate how
these semantic user profiles can be applied in a number of use cases,
including article ranking for personalized search and finding scientists
competent in a topic – e.g., to find reviewers for a paper.

1 Introduction

Researchers increasingly leverage intelligent information systems for managing
their research objects, like datasets, publications, or projects. An ongoing challenge is the overload scientists face when trying to identify potentially relevant
information, e.g., through a web-based search engine: While it is easy to find
numerous potentially relevant results, evaluating each of these is still performed
manually and thus very time-consuming.

We argue that smarter scholarly applications require not just a semantically
rich representation of research objects, but also of their users: By understanding
a scientist's interests, competences, projects and tasks, intelligent systems can

© Springer International Publishing AG 2016
A. González-Beltrán et al. (Eds.): SAVE-SD 2016, LNCS 9792, pp. 113–130, 2016.
DOI: 10.1007/978-3-319-53637-8_12

deliver improved results, e.g., by filtering and ranking results through personalization algorithms [26].

So-called *user profiles* [11,15] have been adopted in domains like e-learning [5], but so far received less attention in scientific applications (we provide a brief background on user profiling in Sect. 2). We believe that a semantically rich representation of users is important for enabling a number of advanced use cases in scholarly applications. We argue that a new generation of *semantic user profile* models are ideally built on standard semantic web technologies, as these make them accessible in an open format to multiple applications that require deeper knowledge of a user's competences and interests. In Sect. 3, we analyze a number of existing Linked Open Data (LOD) [13] vocabularies for describing scholars' preferences and competences. However, they all fall short when it comes to modeling a user's varying degrees of competence in different research topics across different projects. We describe our solution for scholarly user models in Sect. 4.

Bootstrapping such a user profile is an infamous issue in recommendation approaches, known as the *cold start* problem, as asking users to manually create possibly hundreds of entries for their profile is not realistic in practice. Our goal is to be able to create an accurate profile of a scientist's *competences*, which we hypothesize can be automatically calculated based on the publications of the user. Towards this end, we developed the first text mining pipeline that analyzes full-text research articles for an author's competences and exports the results in linked data format into a user profile. The design and implementation of our approach are detailed in Sects. 4 and 5, respectively.

To evaluate our profile generation approach, we performed a user study with ten scientists from two research groups (one in Germany, one in Canada). The participants were provided with two different user profiles each, which were automatically generated based on their publications: One based on the articles' full texts, the second restricted to rhetorical entities (REs) [23]. We asked each participant to rate the relevance of the top-N entries, as well as their competence level. The results, provided in Sect. 6, show that our approach can automatically generate user profiles with a precision of up to 92%.

Finally, we illustrate in Sect. 7 how semantic user profiles can be leveraged by scholarly information systems in a number of use cases, including a competence analysis for a user (e.g., for finding reviewers for a new paper) and re-ranking of article search results, based on a user's profile.[1]

2   Background

In this section, we provide background information on user profiling and its applications. We also briefly introduce semantic technologies for user profiling and their connections with natural language processing (NLP) techniques.

---

[1] For supplementary material, please visit http://www.semanticsoftware.info/save-sd2016.

2.1   User Profiling and Personalization

A user profile is an instance of a user model that contains either a user's characteristics, such as knowledge, interests and backgrounds, or may focus on the context of a user's work, e.g., location and time [5]. Depending on the application offering personalized content, different features have to be taken into account. For instance, educational learning systems typically model a user's knowledge and background, whereas recommender systems and search applications are more focused on a user's interests. Constructing user profiles requires collecting user information over an extended period of time. This gathering process is called *user profiling* and distinguishes between *explicit* and *implicit* user feedback. Explicit user feedback actively requests interests from a user, whereas implicit user feedback derives preferences from user activities. Commonly used implicit profiling techniques, such as extracting preferences from visited websites and deriving interest weights from the numbers of clicks, are discussed by Gauch et al. [11].

User profiles are the basis for a variety of personalized applications. For instance, recommender systems and personalized news portals utilize user information, specifically purchased articles or search terms, in order to adapt content to user needs. The most dominant representation of user characteristics is a weighted vector of keywords, which is still used in many current adaptive web systems [1,17]. This mathematical description makes it possible to apply classical information filtering algorithms, such as cosine similarity [18], in order to measure item-to-item, user-to-user and item-to-user similarity.

2.2   Semantic Technologies

Semantic technologies have become increasingly important in the management of research objects. They allow automated systems to understand the meaning (semantics) and infer additional knowledge from published documents and data [2,25]. Essential building blocks for the creation of structured, meaningful web content are information extraction and semantic annotations – results that can be obtained from NLP pipelines, for example, to detect rhetorical zones, such as *claims* or *contributions* of a paper [23].

In the area of user modeling, a multitude of semantic approaches have emerged in the last decade that use concepts of domain ontologies in the vector representation, rather than keywords [6,26]. In addition to a common understanding of domain knowledge, using semantic technologies also fosters evolving towards more generic user models. A goal of generic user modeling is facilitating software development and promoting reusability [15]. Semantic web technologies, such as the representation of user characteristics in an RDF or OWL format, can leverage this idea. In the following section, we introduce different proposals for generic user modeling with semantic web models. Furthermore, we discuss scholarly ontologies that describe users, institutions and publications in the scientific domain.

3   Literature Review

We focus our review on two core aspects: Firstly, existing semantic vocabularies that describe scholars in academic institutions with their publications and competences, in order to establish semantic user profiles. And secondly, we examine existing approaches for automatic profile generation through NLP methods.

3.1   Vocabularies for Semantic User Profiles

GUMO [14] was the first generic user model approach, designed as a top-level ontology for universal use. This OWL-based ontology focuses on describing a user in a situational context, offering several classes for modeling a user's personality, characteristics and interests. Background knowledge and competences are considered only to a small degree. In contrast, the IntelLEO[2] ontology framework is strongly focused on personalization and enables describing preferences, tasks and interests. The framework consists of multiple RDFS-based ontologies, including vocabularies for user and team modelling, as well as competences. They are inter-linked and can be connected with other user model ontologies, such as FOAF.[3] Due to its simplicity and linkage to other Linked Open Vocabularies, FOAF has become very popular in recent years and is used in numerous personalized applications [7,20,22]. This RDF-based vocabulary permits describing basic user information with predefined entities, such as name, email, homepage, and interests, as well as modeling persons and groups in social networks. However, FOAF does not provide comprehensive classes for describing preferences and competences. Other ontologies attempting to unify user modeling in semantic web applications are the Scrutable User Modelling Infrastructure (SUMI) [16], the Generic User Model Component (GUC) [27] and the ontology developed by Golemati et al. [12].

For modeling scholars in the scientific domain, VIVO[4] [3] is the most prominent approach and has been used in numerous applications.[5] It is an open-source suite of web applications and ontologies used to model scholarly activities across an academic institution. However, VIVO does not provide for content customization, due to missing classes for user interests, preferences and competences. Further vocabularies modeling scientists and publications in research communities are SWRC,[6] SWPO[7] and LSC.[8]

---

[2] IntelLEO (Intelligent Learning Extended Organizations), http://intelleo.eu/index.php?id=183.

[3] FOAF (Friend of a Friend), http://www.foaf-project.org/.

[4] VIVO Ontology, http://vivoweb.org/ontology/core#.

[5] VIVO Registry, http://duraspace.org/registry/vivo.

[6] Semantic Web for Research Communities, http://ontoware.org/swrc/.

[7] Semantic Web Portal Ontology, http://sw-portal.deri.org/ontologies/swportal#.

[8] Linked Science Core Vocabulary, http://linkedscience.org/lsc/ns#.

3.2   Automatic Profile Generation

Generic user models require thinking about new methods for user profiling. Complex user information can be obtained from, e.g., observing a user's browsing behavior, but also from other sources related to the user. Utilizing NLP techniques in user modeling has quite a long history [28]; However, natural language systems are still rarely used for constructing semantic user profiles.

Paik et al. [21] developed <!metaMarker>, an NLP and machine learning pipeline that detects user information in emails. The mined data is used for constructing client profiles in personalized e-commerce applications. The system is able to extract explicit metadata, such as 'name of sender', 'title' or 'affiliation', as well as implicit metadata, like 'mood' or 'intention' of the user. Additionally, they enriched this context-related metadata with new elements, such as 'like', 'dislike', 'interested' and 'not interested', in order to describe a user's preferences. The pipeline consists of seven steps, including Sentence Splitting, Part-Of-Speech Tagging, Stemming and Entity Extraction, generating explicit user information at the end. Through Bayesian probabilistic and k-Nearest Neighbour classifiers, mood and intentions are determined. A training set of 5000 emails was used to build the classifiers for the implicit metadata. The effectiveness of the system was measured with precision and recall, resulting in an average precision of 89%.

LinkedVis [4] is an interactive recommender system that generates career recommendations and supports users in finding potentially interesting companies and specific roles. The authors designed four different user models based on data from *LinkedIn*[9] and extracted interests and preferences from a user's connections, average roles and companies. Two of the four constructed profiles contained meaningful entities instead of plain keywords. A Part-of-Speech Tagger was utilized to find noun phrases that were mapped to Wikipedia articles. The evaluation with a leave-one-out cross-validation revealed that the user models with the semantic enrichment produced more accurate and more diverse recommendations than the profiles based on TF-IDF weights and occurrence matching.

Another approach using NLP methods for online profile resolution is proposed by Cortis et al. [8]. They developed a system for analyzing user profiles from heterogeneous online resources in order to aggregate them into one unique profile. For this task, they used GATE's ANNIE[10] plugin [9] and adapted its JAPE grammar rules to disassemble a person's name into five sub-entities such as prefix, suffix, first name, middle name and surname. In addition, a Large Knowledge Base (LKB) Gazetteer was incorporated to extract supplementary city and country values from DBpedia.[11] In their approach, location-related attributes (e.g., Dublin and Ireland) could be linked to each other based on these semantic extensions, where a string-matching approach would have failed. In their user evaluation, the participants were asked to assess their merged profile on a binary rating scale. More than 80% of the produced profile entries were marked as correct. The results reveal that profile matchers can improve the

---

[9] LinkedIn, https://www.linkedin.com.

[10] ANNIE, https://gate.ac.uk/sale/tao/splitch6.html.

[11] DBpedia, http://dbpedia.org.

management of one's personal information across different social networks and support recommendations of possibly interesting new contacts based on similar preferences.

3.3 Discussion

As presented above, there exist only few automatic user profiling approaches using linked named entities and NLP techniques. The most widespread description of a user model in these applications is still a term-based vector representation. Even though keywords are increasingly replaced by linked entities, they still lack an underlying semantic model in RDF or OWL format. With respect to existing application domains, social networks are common sources for gathering personal information. Scholars in particular were not considered in any of the aforementioned systems.

In contrast, we aim at automatically creating semantic user profiles for scholars by means of NLP methods and semantic web technologies. Our goal is to establish user profiles in an RDF format that can be stored in a triplestore. Hosting user information in a structured and meaningful semantic format facilitates data integration across different sources. Furthermore, expressive SPARQL queries and inferences can help to discover related preferences that are not explicitly stated in the profiles.

4 Design

In our approach, we take the publications of an author as input to an automated text mining pipeline, which creates user profiles in LOD format, based on the competences detected in the papers. The hypothesis behind our design is that authors of a scholarly publication (e.g., a journal article) are competent in the topics mentioned in the paper to various degrees. Our text mining system performs entity linking from scholarly documents and generates competence relations between a document's authors and its contained LOD named entities using linked open vocabularies. The result is a knowledge base containing the semantic profiles of authors that can be exploited for a variety of use cases by humans and machines alike, as we show in Sect. 7.

4.1 Semantic Modeling of Users' Competence Records

Modeling semantic scholarly profiles requires the formalization of the relation between authors, their publications, and the topics mentioned in them in a semantically rich and interoperable format. To this end, we decided to use the W3C standard RDF framework to design profiles based on semantic triples. Since RDF documents intrinsically represent labeled, directed graphs, the semantic profiles of scholars extracted from the documents can be merged through common competence URIs, i.e., authors extracted from otherwise disparate documents can be semantically related using their competence topics.

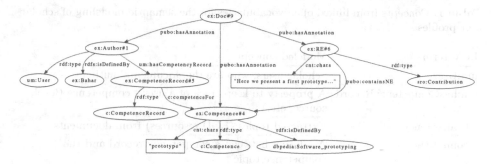

Fig. 1. A semantic scholar profile in form of an RDF graph

Following the best practices of producing linked open datasets, we tried to reuse existing Linked Open Vocabularies (LOVs) to the extent possible for modeling the extracted knowledge. Table 1 shows the vocabularies used to model our semantic scholarly profiles. We largely reuse IntelLEO ontologies for competence modeling – originally designed for semantic modeling of learning contexts –, in particular the vocabularies for *User and Team Modeling*[12] and *Competence Management*.[13] We also reuse the PUBO ontology [23] for modeling the relation between the documents that we process, the generated annotations and their inter-relationships. Figure 1 shows a minimal example semantic profile in form of an RDF graph.

4.2   Automatic Detection of Competences

Our text mining system accepts a set of publications from an author as input and processes the full-text of the documents to detect competence topics, i.e., grounded Named Entities (NEs). Each document first goes through a pre-processing phase. In this phase, the full-text of the document is segmented into tokens: smaller, linguistically meaningful parts, like words, numbers and symbols. Subsequent syntactical processing components process the tokenized text into sentences and all sentence constituents are tagged with a Part-of-Speech category. Grammatical processing of sentences helps us to filter out the text tokens that do not represent competences, like adverbs or pronouns. Lastly, we ground (link) nouns and noun phrases in text to their corresponding resource (sense) in the LOD cloud. To this end, we selected the DBpedia Spotlight [19] annotation tool that can link the surface forms of terms in a document to a URI in the DBpedia ontology that serves as the nucleus of the LOD cloud. In this paper, we use the raw frequency of these NEs in documents as a means of ranking the top competence topics for researchers' profiles. Finally, once the documents are processed, we go over the generated annotations and transform them into RDF triples, using the vocabularies described in Sect. 4.1.

---

[12] IntelLEO User Model Ontology, http://intelleo.eu/ontologies/user-model/spec.

[13] IntelLEO Competence Ontology, http://www.intelleo.eu/ontologies/competences/spec.

Table 1. Concepts from linked open vocabularies for the semantic modeling of scholar user profiles

LOV term	Modeled concept
um:User	Scholar users, who are the documents' authors
um:hasCompetencyRecord	A property to keep track of a user's competence (level, source, etc.)
c:Competency	Extracted topics (LOD resources) from documents
c:competenceFor	A relation between a competency record and the competence topic
sro:RhetoricalElement	A sentence containing a rhetorical entity, e.g., a *contribution*
cnt:chars	A competence's label (surface form) as appeared in the document
pubo:hasAnnotation	A property to relate annotations to documents
pubo:containsNE	A property to relate rhetorical zones and entities in the document
oa:start & oa:end	A property to show the start/end offsets of competences in text

um:	http://intelleo.eu/ontologies/user-model/ns/
c:	http://intelleo.eu/ontologies/competences/ns/
sro:	http://salt.semanticauthoring.org/ontologies/sro#
cnt:	http://www.w3.org/2011/content#
pubo:	http://lod.semanticsoftware.info/pubo/pubo#
oa:	http://www.w3.org/ns/oa/
rdf:	http://www.w3.org/1999/02/22-rdf-syntax-ns#
rdfs:	http://www.w3.org/2000/01/rdf-schema#

A rather interesting question here is whether all of the detected entities are representative of the authors' interest, or if topics in certain regions of the documents are better candidates? To test this hypothesis, we further process the documents to annotate their so-called *Rhetorical Entities (REs)*, where authors convey their findings in form of claims or arguments, by looking at their linguistic features [24]. In this fashion, we can later evaluate whether the NEs in RE zones of documents better represent the authors' competences.

5   Implementation

In this section, we describe how we realized the semantic user profiling of authors illustrated in the previous section.

5.1   Extraction of User Competences with Text Mining

We developed a text mining pipeline, implemented based on the GATE framework, to analyze a given author's papers to automatically extract the competence records and topics. The NLP pipeline accepts a corpus (set of documents) for each author as input. We use GATE's ANNIE plugin to pre-process each document's full-text and further process all sentences with a Part-of-Speech (POS) tagger, so that their constituents are labeled with a POS tag, such as *noun, verb,* or *adjective* and lemmatized to their canonical (root) form. We use MuNPEx,[14] a GATE plugin to detect noun phrases in text, which helps us to extract competence topics that are noun phrases rather than nouns alone. Subsequently, we use our LODtagger,[15] which is a GATE plugin that acts as a wrapper for the annotation of documents with Named Entity Recognition tools. In our experiments, we use a local installation of DBpedia Spotlight v7.0 with a statistical model[16] for English [10]. Spotlight matches the surface form of the document's tokens against the DBpedia ontology and links them to their corresponding resource URI. LODtagger then transforms the Spotlight response to GATE annotations using the entities' offsets in text and keeps their URI in the annotation's features.

To evaluate whether our hypothesis that the NEs within rhetorical zones of a document are more representative of the author's competences than the NEs that appear anywhere in the document, we decided to annotate the *Claim* and *Contribution* sentences of the documents using our Rhetector[17] GATE plugin [23]. This way, we can create user profiles exclusively from the competence topics that appear within these RE annotations for comparison against profiles populated from full-text.[18] Finally, we create a competence record between the author and each of the detected competences (represented as DBpedia NEs). We use GATE's JAPE language that allows us to execute regular expressions over documents' annotations by internally transforming them into finite-state machines. Thereby, we create a competence record (essentially, a GATE relation) between the author annotation and every competence topic in the document.

5.2 Automatic Population of Semantic User Profiles

The last step in our automatic generation of semantic user profiles is to export all of the GATE annotations and relations from the syntactic and semantic processing phases into semantic triples using RDF. Our LODeXporter[19] tool provides a flexible mapping of GATE annotations to RDF triples with user-defined transformation rules. For example, the rules:

```
map:GATECompetence    map:GATEtype    "DBpediaNE".
map:GATECompetence    map:hasMapping    map:GATELODRefFeatureMapping.
map:GATELODRefFeatureMapping    map:GATEfeature    "URI".
map:GATELODRefFeatureMapping    map:type    rdfs:isDefinedBy.
```

---

[14] Multi-lingual Noun Phrase Extractor (MuNPEx), http://www.semanticsoftware.info/munpex.

[15] LODtagger, http://www.semanticsoftware.info/lodtagger.

[16] DBpedia statistical model for English (en_2+2), http://spotlight.sztaki.hu/downloads/.

[17] Rhetector, http://www.semanticsoftware.info/rhetector.

[18] Rhetector was evaluated in [23] with an average F-measure of 73%.

[19] LODeXporter, http://www.semanticsoftware.info/lodexporter.

SEMANTIC USER PROFILING EVALUATION SHEET

NAME: _____

For each topic, please choose *only one* of the available options that best represents your level of expertise:

Novice means "*I am somewhat familiar with this topic.*"
Intermediate means "*I have conducted research on this topic and feel competent in it.*"
Advanced means "*I am an expert in this topic.*"

#	Competency Topic	Novice	Intermediate	Advanced	Irrelevant
1	Recommender system	☐	☐	☐	☐

Recommender systems or recommendation systems (sometimes replacing "system" with a synonym such as platform or engine) are a subclass of information filtering system that seek to predict the 'rating' or 'preference' that user would give to an item. Recommender systems have become extremely common in recent years, and are applied in a variety of applications. The most popular ones are probably movies, music, news, books, research articles, search queries, social tags, and products in general.

Fig. 2. Excerpt of a sample generated user profile for evaluation

describe that all "*DBpediaNE*" annotations in the document should be exported, and for each annotation the value of its "*URI*" feature can be used as the object of the triple, using "*rdfs:isDefinedBy*" as the predicate. Similarly, we use the LOV terms shown in Table 1 to model authors, competence records and topics as semantic triples and store the results in an Apache TDB-based[20] triplestore.

6 Evaluation

To evaluate the accuracy of the generated profiles, we reached out to ten computer scientists from Concordia University and the University of Jena (including the authors of this paper) and asked them to provide us with a number of their selected publications. We processed the documents and populated a knowledge base with the researchers' profiles. We also developed a Java command-line tool that queries the knowledge base and generates LaTeX documents to provide for a human-readable format of the researchers' profiles (shown in Fig. 2) that lists their top-50 competence topics sorted by the number of occurrence in the users' publications. Subsequently, we asked the researchers to review their profiles across two dimensions: *(i)* relevance of the extracted competences, and *(ii)* their level of expertise for each extracted competence.

For each participant, we exported two versions of their profile: *(i)* a version with a list of competences extracted from their papers' full-text, and *(ii)* a second version that only lists the competences extracted from the rhetorical zones of the documents, in order to test our hypothesis described in Sect. 5.1. To ensure that none of the competence topics are ambiguous to the participants, our command-line tool also retrieves the English label and comment of each topic from the DBpedia ontology using its public SPARQL endpoint.[21] The participants were

---

[20] Apache TDB, http://jena.apache.org/documentation/tdb/.
[21] DBpedia public SPARQL endpoint, http://dbpedia.org/sparql.

instructed to choose only one level of expertise for each competence and choose *"irrelevant"* if the competence topic was incorrect or grounded to a wrong sense.

To evaluate the effectiveness of our system, we utilize one of the most popular ranked retrieval evaluation methods, namely the Mean Average Precision (MAP) [18]. MAP indicates how precise an algorithm or system ranks its top-N results, assuming that the entries listed on top are more relevant for the information seeker than the lower ranked results. Table 2 shows the evaluation results of our user study. A competence was considered as relevant when it had been assigned to one of the three levels of expertise (novice, intermediate, advanced). For each participant, we measured the average precision of the generated profiles in both the full-text and RE-only versions. Here, precision is evaluated at a given cut-off rank N, considering only the top-N results returned by the system. Hence, MAP is the mean of the average precisions at each cut-off rank. The results show that for both the top-10 and top-25 competences, 70–80% of the profiles generated from RE-only zones had a higher precision, increasing the system MAP up to 4% in each cut-off. In the top-50 column, we observed a slight decline in some of the profiles' average precision, which we believe to be a consequence of more irrelevant topics appearing in the profiles, although the MAP score stays almost the same for both versions. Analyzing the distribution of answers across the three levels of expertise, the results illustrated in Fig. 3 reveal that in both versions, around 60% of the detected competences are related to either the intermediate or advanced level.

Finally, all participants (except R10) informally stated that the RE-only version of their profiles were better representing their competences, corroborating

Table 2. Evaluation of the generated user profiles

Participant	#Docs	#Distinct Competences		Avg. Precision@10		Avg. Precision@25		Avg. Precision@50	
		Full Doc	REs Only	Full Doc	REs Only	Full Doc	REs Only	Full Doc	REs Only
R1	8	2,718	293	**0.91**	0.80	**0.84**	0.74	**0.80**	0.69
R2	7	2,096	386	**0.95**	0.91	0.90	**0.92**	0.87	**0.91**
R3	6	1,200	76	0.96	**0.99**	0.93	**0.95**	**0.92**	0.88
R4	5	1,240	149	0.92	**0.92**	**0.86**	0.81	**0.77**	0.75
R5	4	1,510	152	0.84	**0.99**	0.87	**0.90**	0.82	**0.82**
R6	6	1,638	166	0.93	**1.0**	0.90	**0.97**	0.88	**0.89**
R7	3	1,006	66	0.70	**0.96**	0.74	**0.89**	0.79	**0.86**
R8	8	2,751	457	0.96	**1.0**	0.92	**1.0**	0.92	**0.99**
R9	9	2,391	227	0.67	**0.73**	0.62	**0.70**	0.56	**0.65**
R10	5	1,908	176	**0.96**	0.91	0.79	**0.80**	0.69	**0.70**
			MAP	0.88	**0.92**	0.83	**0.87**	0.80	**0.81**

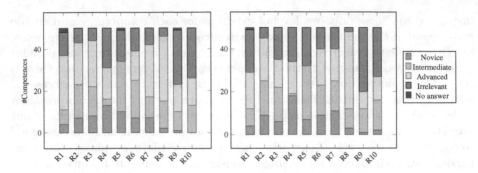

Fig. 3. Distribution of competence levels in full-text (left) and RE-only (right) profiles

our hypothesis that the topics mentioned in the RE zones of a document are more accurate in representing its authors' competences. This is encouraging because, as shown in Table 2, compared to the number of distinct competences extracted from the full-text of documents, we need an order of a magnitude fewer topics, which not only better represent the users' competences, but also significantly reduces the size of the knowledge base.

7   Application

In this section, we demonstrate a number of use cases in which semantic user profiles can play an effective role.

7.1   Finding All Competences of a User

By querying the populated knowledge base with the researchers' profiles, we can find all topics that a user is competent in. Following our knowledge base schema (see Sect. 4), we can query all the competence records of a given author URI and find the topics (in form of LOD URIs), from either the papers' full-text or exclusively the RE zones. In fact, the SPARQL query shown below is how we gathered each user's competences (from RE zones) to generate the evaluation profiles described in Sect. 6:

```
SELECT DISTINCT ?uri (COUNT(?uri) AS ?count) WHERE {
    ?creator rdf:type um:User .
    ?creator rdfs:isDefinedBy <http://semanticsoftware.info/lodexporter/creator/R1> .
    ?creator um:hasCompetencyRecord ?competenceRecord .
    ?competenceRecord c:competenceFor ?competence .
    ?competence rdfs:isDefinedBy ?uri .
    ?rhetoricalEntity rdf:type sro:RhetoricalElement .
    ?rhetoricalEntity pubo:containsNE ?competence .
} GROUP BY ?uri ORDER BY DESC(?count)
```

Table 3 shows a number of competence topics (grounded to their LOD URIs) for some of our evaluation participants, sorted in descending order by their frequency in the documents.

Table 3. A number of users and their most frequent competence topics

User	Extracted Competence Topics
R1	dbpedia:Tree_(data_structure), dbpedia:Vertex_(graph_theory), dbpedia:Cluster_analysis, ...
R2	dbpedia:Natural_language_processing, dbpedia:Semantic_Web, dbpedia:Entity-relationship_model, ...
R3	dbpedia:Recommender_system, dbpedia:Semantic_web, dbpedia:Web_portal, dbpedia:Biodiversity, ...
R4	dbpedia:Service_(economics), dbpedia:Feedback, dbpedia:User_(computing), dbpedia:System, ...
R5	dbpedia:Result, dbpedia:Service_discovery, dbpedia:Web_search_engine, dbpedia:Internet_protocol, ...

7.2 Ranking Papers Based on a User's Competences

Semantic user profiles can be incredibly effective in the context of information retrieval systems. Here, we demonstrate how they can help to improve the relevance of the results. Our proposition is that papers that mention the competence topics of a user are more *interesting* for her and thus, should be ranked higher in the results. Therefore, the diversity and frequency of topics within a paper should be used as ranking features. We showed in [23] that retrieving papers based on their LOD entities is more effective than conventional keyword-based methods. However, the results were not presented in order of their *interestingness* for the end-user. Here, we integrate our semantic user profiles to re-rank the results, based on the common topics in both the papers and a user's profile:

```
SELECT (COUNT(DISTINCT ?uri) as ?rank) WHERE {
    <http://example.com/example_paper.xml> pubo:hasAnnotation ?topic .
    ?topic  rdf:type  pubo:LinkedNamedEntity .
    ?topic  rdfs:isDefinedBy  ?uri .
    FILTER EXISTS {
        ?creator  rdfs:isDefinedBy  <http://semanticsoftware.info/lodexporter/creator/R8> .
        ?creator  um:hasCompetencyRecord ?competenceRecord .
        ?competenceRecord c:competenceFor ?competence .
        ?competence rdfs:isDefinedBy ?uri .} }
```

The query shown above compares the topic URIs in a given paper to user R8's competences extracted from full-text documents and counts the occurrence of such a hit. Note that the DISTINCT keyword will cause the query to only count the unique topics, e.g., if <dbpedia:Semantic_Web> appears two times in the paper, it will be counted as one occurrence.[22] We can then use the numbers returned by the query above as a means to rank the papers. Table 4 shows the result set returned by performing a query against the SePublica dataset of 29 papers from [23] to find papers mentioning <dbpedia:Ontology_(information_science)>.

---

[22] We decided to the count the unique occurrences, because a ranking algorithm based on the raw frequency of competence topics will favour long (non-normalized) papers over shorter ones.

Table 4. Re-ranking of the top-10 search results, originally sorted by a frequency-based method, through integrating semantic user profiles

Paper Title	Topic Mentions		R8's Profile		R6's Profile	
	Rank	Raw Frequency	Rank	Com. Topics	Rank	Com. Topics
A Review of Ontologies for Describing Scholarly and Scientific Documents	1	92	1	312	5	198
BauDenkMalNetz - Creating a Semantically Annotated Web Resource of Historical Buildings	2	50	5	294	4	203
Describing bibliographic references in RDF	3	38	6	269	8	177
Semantic Publishing of Knowledge about Amino Acids	4	25	10	79	10	53
Supporting Information Sharing for Re-Use and Analysis of Scientific Research Publication Data	5	25	4	306	7	185
Linked Data for the Natural Sciences: Two Use Cases in Chemistry and Biology	6	23	2	310	1	220
Ornithology Based on Linking Bird Observations with Weather Data	7	22	8	248	6	189
Systematic Reviews as an Interface to the Web of (Trial) Data: using PICO as an Ontology for Knowledge Synthesis in Evidence-based Healthcare Research	8	19	9	179	9	140
Towards the Automatic Identification of the Nature of Citations	9	19	3	307	2	214
SMART Research using Linked Data - Sharing Research Data for Integrated Water Resources Management in the Lower Jordan Valley	10	19	7	260	3	214

The "*Topic Mentions*" column shows the ranked results based on how many times the query topic was mentioned in a document. In contrast, the R6 and R8profile-based columns show the ranked results using the number of common topics between the papers (full-text) and the researchers' respective profiles (populated from full-text documents). Note that in the R6 and R8profile-based columns, we only count the number of unique topics and not their frequency.

An interesting observation here is that the paper ranked fourth in the frequency-based column ranks last in both profile-based result sets. A manual inspection of the paper revealed that this document, although originally ranked high in the results, is in fact an editors' note in the preface of the SePublica 2012 proceedings. On the other hand, the paper which ranked first in the frequency-based column, remained first in R8's result set, since he has a stronger research focus on ontologies and linked open data compared to R6, as we observed from their generated profiles during evaluation.

7.3   Finding Users with Related Competences

Given the semantic user profiles and a topic in form of an LOD URI, we can find all users in the knowledge base that have related competences. By virtue of traversing the LOD cloud, we can find topic URIs that are (semantically) related to a given competence topic and match against users' profiles to find competent authors:

```
PREFIX dcterms: <http://purl.org/dc/terms/>
PREFIX dbpedia: <http://dbpedia.org/resource/>

SELECT ?author_uri WHERE {
    SERVICE <http://dbpedia.org/sparql> {
        dbpedia:Ontology_( information_science ) dcterms:subject ?category .
        ?subject    dcterms:subject ?category . }
?author rdf:type um:User .
?creator rdfs:isDefinedBy ? author_uri .
?creator um:hasCompetencyRecord ?competenceRecord.
?competenceRecord c:competenceFor ?competence.
?competence rdfs:isDefinedBy ?subject .
? rhetoricalEntity pubo:containsNE ?competence.
? rhetoricalEntity rdf:type sro:RhetoricalElement . }
```

The query above first performs a federated query against DBpedia's SPARQL endpoint to find topic URIs that are semantically related to the query topic.[23] Then, it matches the retrieved URIs against the topics of the knowledge base users' competence records. This way, for example as shown in Table 5, even if

Table 5. Topics related to the query and their respective competent researchers

Competence Topic	Competent Users
dbpedia:Ontology_(information_science)	R1, R2, R3, R8
dbpedia:Linked_data	R2, R3, R8
dbpedia:Knowledge_representation_and _reasoning	R1, R2, R4, R8
dbpedia:Semantic_Web	R1, R2, R3, R4, R5, R6, R7, R8
dbpedia:Controller_vocabulary	R2, R3, R8
dbpedia:Tree_(data_structure)	R1, R4, R7

---

[23] We assume all topics under the same category in the DBpedia ontology are semantically related.

a researcher does not have <dbpedia:Ontology_(information_science)>, but does have <dbpedia:Linked_data> in her profile, she will be returned as a hit, since both of the aforementioned topics are related in the DBpedia ontology. In other words, if we are looking for persons competent in ontologies, a researcher that has previously conducted research on linked data might also be a suitable match.

8 Conclusions

Semantic user profiles are an important extension for semantic publishing applications: With a standardized, shareable, and extendable representation of a user's competences, a number of novel scenarios become possible. Searching for scientists with specific competences can help to find reviewers for a given paper or proposal. Recommendation algorithms can filter and rank the immense amount of research objects, based on the profile of individual users. And a wealth of additional applications becomes feasible, such as matching the competences of a research group against project requirements, simply by virtue of analyzing an inter-linked knowledge graph of users, datasets, publications, and other artifacts. The work presented here demonstrates how we can represent scholarly profiles in LOD format. We show how to bootstrap semantic user profiles including scientists' competences through an automated text mining approach with high accuracy. In ongoing work, we are currently integrating the semantic user profiles into a scholarly data portal for biodiversity research, in order to evaluate their impact on concrete research questions in a life sciences scenario.

Acknowledgments. We would like to thank all the participants in our user study.

References

1. Almuhaimeed, A., Fasli, M.: A semantic method for multiple resources exploitation. In: Proceedings of the 11th International Conference on Semantic Systems (SEMANTICS 2015), pp. 113–120. ACM, New York (2015)
2. Berners-Lee, T., Hendler, J.: Publishing on the semantic web. Nature **410**, 1023–1024 (2001)
3. Börner, K., Conlon, M., Corson-Rikert, J., Ding, Y.: VIVO: A Semantic Approach to Scholarly Networking and Discovery. Synthesis Lectures on the Semantic Web. Morgan & Claypool Publishers, San Rafael (2012)
4. Bostandjiev, S., O'Donovan, J., Höllerer, T.: Linkedvis: exploring social and semantic career recommendations. In: Proceedings of the 2013 International Conference on Intelligent User Interfaces (IUI 2013), pp. 107–116. ACM, New York (2013)
5. Brusilovsky, P., Millán, E.: User models for adaptive hypermedia and adaptive educational systems. In: Brusilovsky, P., Kobsa, A., Nejdl, W. (eds.) The Adaptive Web. LNCS, vol. 4321, pp. 3–53. Springer, Heidelberg (2007). doi:10.1007/978-3-540-72079-9_1
6. Cantador, I., Castells, P.: Extracting multilayered communities of interest from semantic user profiles: application to group modeling and hybrid recommendations. Comput. Hum. Behav. **27**(4), 1321–1336 (2011)

7. Celma, Ò.: Foafing the music: bridging the semantic gap in music recommendation. In: Cruz, I., Decker, S., Allemang, D., Preist, C., Schwabe, D., Mika, P., Uschold, M., Aroyo, L.M. (eds.) ISWC 2006. LNCS, vol. 4273, pp. 927–934. Springer, Heidelberg (2006). doi:10.1007/11926078_67
8. Cortis, K., Scerri, S., Rivera, I., Handschuh, S.: An ontology-based technique for online profile resolution. In: Jatowt, A., Lim, E.-P., Ding, Y., Miura, A., Tezuka, T., Dias, G., Tanaka, K., Flanagin, A., Dai, B.T. (eds.) SocInfo 2013. LNCS, vol. 8238, pp. 284–298. Springer, Cham (2013). doi:10.1007/978-3-319-03260-3_25
9. Cunningham, H., et al.: Text Processing with GATE (Version 6). University of Sheffield, Department of Computer Science (2011). http://tinyurl.com/gatebook
10. Daiber, J., Jakob, M., Hokamp, C., Mendes, P.N.: Improving efficiency and accuracy in multilingual entity extraction. In: Proceedings of the 9th International Conference on Semantic Systems (I-Semantics) (2013)
11. Gauch, S., Speretta, M., Chandramouli, A., Micarelli, A.: User profiles for personalized information access. In: Brusilovsky, P., Kobsa, A., Nejdl, W. (eds.) The Adaptive Web. LNCS, vol. 4321, pp. 54–89. Springer, Heidelberg (2007). doi:10.1007/978-3-540-72079-9_2
12. Golemati, M., Katifori, A., Vassilakis, C., Lepouras, G., Halatsis, C.: Creating an ontology for the user profile: method and applications. In: Proceedings of the First International Conference on Research Challenges in Information Science (RCIS) (2007)
13. Heath, T., Bizer, C.: Linked Data: Evolving the Web into a Global Data Space. Synthesis lectures on the semantic web: theory and technology. Morgan & Claypool Publishers, San Rafael (2011)
14. Heckmann, D., Schwartz, T., Brandherm, B., Schmitz, M., Wilamowitz-Moellendorff, M.: GUMO – the general user model ontology. In: Ardissono, L., Brna, P., Mitrovic, A. (eds.) UM 2005. LNCS (LNAI), vol. 3538, pp. 428–432. Springer, Heidelberg (2005). doi:10.1007/11527886_58
15. Kobsa, A.: Generic user modeling systems. User Model. User Adapted Interact. 11(1–2), 49–63 (2001)
16. Kyriacou, D., Davis, H.C., Tiropanis, T.: A (multi'domain'sional) scrutable user modelling infrastructure for enriching lifelong user modelling. In: Lifelong User Modelling Workshop (in Conjunction with Conference UMAP 2009), Trento, Italy (2009)
17. Malhotra, A., Totti, L.C., Jr., W.M., Kumaraguru, P., Almeida, V.: Studying user footprints in different online social networks. CoRR abs/1301.6870 (2013)
18. Manning, C.D., Raghavan, P., Schütze, H.: Introduction to Information Retrieval. Cambridge University Press, Cambridge (2008)
19. Mendes, P.N., Jakob, M., García-Silva, A., Bizer, C.: DBpedia Spotlight: shedding light on the web of documents. In: Proceedings of the 7th International Conference on Semantic Systems, pp. 1–8. ACM (2011)
20. Orlandi, F., Breslin, J., Passant, A.: Aggregated, interoperable and multi-domain user profiles for the social web. In: Proceedings of the 8th International Conference on Semantic Systems (I-SEMANTICS 2012), pp. 41–48. ACM, New York (2012)
21. Paik, W., Yilmazel, S., Brown, E., Poulin, M., Dubon, S., Amice, C.: Applying natural language processing (NLP) based metadata extraction to automatically acquire user preferences. In: Proceedings of the 1st International Conference on Knowledge Capture (K-CAP 2001), pp. 116–122. ACM, New York (2001)
22. Raad, E., Chbeir, R., Dipanda, A.: User profile matching in social networks. In: The 13th International Conference on Network-Based Information System (2010)

23. Sateli, B., Witte, R.: Semantic representation of scientific literature: bringing claims, contributions and named entities onto the Linked Open Data cloud. PeerJ Comput. Sci. **1**, e37 (2015). https://peerj.com/articles/cs-37/
24. Sateli, B., Witte, R.: What's in this paper? Combining rhetorical entities with linked open data for semantic literature querying. In: Semantics, Analytics, Visualisation: Enhancing Scholarly Data (SAVE-SD 2015), pp. 1023–1028. ACM, Florence, Italy (2015). http://www.www2015.it/documents/proceedings/companion/p1023.pdf
25. Shadbolt, N., Hall, W., Berners-Lee, T.: The semantic web revisited. IEEE Intell. Syst. **21**(3), 96–101 (2006)
26. Sieg, A., Mobasher, B., Burke, R.: Web Search Personalization with ontologicaluser profiles. In: Proceedings of the 16th ACM Conference on Information and Knowledge Management (CIKM 2007), pp. 525–534. ACM, New York (2007)
27. van der Sluijs, K., Houben, G.J.: Towards a generic user model component. In: Workshop on Personalization on the Semantic Web (PerSWeb 2005), Edinburgh, Scotland, pp. 47–57 (2005)
28. Zukerman, I., Litman, D.: Natural language processing and user modeling: synergies and limitations. User Model. User Adapted Interact. **11**(1–2), 129–158 (2001)

Detection of Embryonic Research Topics by Analysing Semantic Topic Networks

Angelo Antonio Salatino$^{(\boxtimes)}$ and Enrico Motta

Knowledge Media Institute, The Open University, Milton Keynes, UK
{angelo.salatino,enrico.motta}@open.ac.uk

Abstract. Being aware of new research topics is an important asset for anybody involved in the research environment, including researchers, academic publishers and institutional funding bodies. In recent years, the amount of scholarly data available on the web has increased steadily, allowing the development of several approaches for detecting emerging research topics and assessing their trends. However, current methods focus on the detection of topics which are already associated with a label or a substantial number of documents. In this paper, we address instead the issue of detecting embryonic topics, which do not possess these characteristics yet. We suggest that it is possible to forecast the emergence of novel research topics even at such early stage and demonstrate that the emergence of a new topic can be anticipated by analysing the dynamics of pre-existing topics. We present an approach to evaluate such dynamics and an experiment on a sample of 3 million research papers, which confirms our hypothesis. In particular, we found that the pace of collaboration in sub-graphs of topics that will give rise to novel topics is significantly higher than the one in the control group.

Keywords: Ontology · Research trend detection · Scholarly data · Semantic web · Topic discovery · Topic emergence detection

1 Introduction

Being aware of new research topics is important for anybody involved in the research environment and, although the effective detection of new research trends is still an open problem, the availability of very large repositories of scholarly data and other relevant sources opens the way to novel data-intensive approaches to address this problem. We can consider two main phases in the early life of a topic. In its *initial stage*, a group of scientists agree on some basic theories, build a conceptual framework and begin to establish a new scientific community. Afterwards, the new area enters a *recognised phase* in which a substantial number of authors start working on it, producing and disseminating results. This characterisation is consistent with Kuhn's vision of scientific revolutions [12]. There are already several approaches capable of detecting novel topics and research trends [4,6,10], which rely on statistical techniques to analyse the impact of either labels or distributions of words associated to topics. However, all these

© Springer International Publishing AG 2016
A. González-Beltrán et al. (Eds.): SAVE-SD 2016, LNCS 9792, pp. 131–146, 2016.
DOI: 10.1007/978-3-319-53637-8_13

approaches are able to recognise topics only in the two aforementioned phases; that is, when they are already established and associated with a substantial number of publications and when the communities of researchers have already reached a consensus for a label. In this paper, we focus on the earlier *embryonic phase*, in which the topic itself has not yet been explicitly labelled or identified by a research community. We theorise that it is possible to detect topics at this stage by analysing the dynamics of existent topics. This hypothesis follows from a number of theories [1, 12, 23] which suggest that new topics actually derive from the interactions and cross-pollinations of established research areas. We present a method which integrates statistics and semantics for assessing the dynamics of a topic graph. The method was tested on a sample of 3 million papers and the experiment confirmed our hypothesis. In particular, it was found that the pace of collaboration in graphs of topics that will give rise to a new topic is significantly higher than the one of the control group. This paper is organised as follows. Section 2 introduces the state of the art. In Sect. 3 we describe the experimental approach used to confirm our hypothesis and in Sect. 4 we show and discuss the results. We conclude in Sect. 5 by discussing the future directions of our research.

2   Related Work

Detecting topics and their trends is a task that has recently gained increased interest from the information retrieval community and has been applied to many contexts, such as social networks [15], blogs [9], emails [16] and scientific literature [4, 5, 8, 14, 24].

The state of the art presents several works on research trend detection, which can be characterised either by the way they define a topic or the techniques they use to detect them [22]. Latent Dirichlet Allocation (LDA) [3] is an unsupervised learning method to extract topics from a corpus and models topics as a multinomial distribution over words. Since its introduction, LDA has been extended and adapted in several applications. For example, He *et al.* [10] combined LDA and citation networks in order to address the problem of topic evolution. Their approach detects topics in independent subsets of a corpus and then leverages citations to connect topics in different time frames. Similarly, Rosen-Zvi *et al.* [21] and Bolelli *et al.* [4] extend LDA with the Author-Topic model, in which authors can shape the distribution of topics, and claim that their approach is capable of detecting more new hidden topics than the standard LDA approach. However, these approaches model topics as a distribution over words making difficult to label them, and also the number of topics need to be known a priori.

Morinaga *et al.* [16] employ the Finite Mixture Model to represent the structure of topics and analyse the changes in time of the extracted components to track emerging topics. This approach was evaluated on an email corpus and therefore is not clear how it could perform on scientific literature, especially when the full text of papers is not available.

Duvvuru *et al.* [6, 7] analysed networks of co-occurring keywords in scholarly articles and monitored the evolution in time of the link weights for detecting research trends and emerging research areas. However, as pointed out by previous works [19], keywords tend to be noisy and do not always represent research topics. For example, Osborne et al. [20] show that the use of a semantic characterisation of research topics yields better results for the detection of research communities.

To alleviate this problem, Decker et al. [5] matched a corpus of publications to a taxonomy of topics based on the most significant words found in titles and abstracts, and analysed the changes in the number of publications associated with topics. Similarly, Erten et al. [8] adopted the ACM Digital Library taxonomy for analysing the evolution of topic graphs to monitor research trends. In our experiment we adopted a similar solution and used an ontology of computer science generated and regularly maintained by the Klink-2 algorithm [18], which has the advantage of being always up to date.

Jo et al. [11] have developed an approach that correlates distributions of terms with the distribution of the citation graph related to publications containing that term. Their work is based on the intuition that if a term is relevant to a particular topic, documents containing that term will have a stronger connection than randomly selected ones. However, this approach is not suitable for emerging topics since it will take time for the citation network of a term to become tightly connected.

To summarise, the state of the art presents several approaches for detecting research trends. However these focus on already recognised topics, associated with a label or, in the case of probabilistic topics models, with a set of terms. Therefore, the problem of detecting research trends in their embryonic phase still needs to be addressed.

3   Experiment Design

In order to confirm the theory that the emergence of a new topic is actually anticipated by the dynamics between already established topics, we designed the following experiment. We selected 50 topics debuting between 2000 and 2010 and extracted the sub-graphs of the n keywords most co-occurring with each topic. We then analysed these graphs in the five years before the topic debut year and compared them to a control group of graphs associated with established topics.

The full list of topics and the results of the experiment can be found at http:// technologies.kmi.open.ac.uk/rexplore/www2016/. In the following sections we will describe the dataset, the steps of the process and the metrics used to measure the pace of collaboration of the sub-graphs.

3.1   Dataset

The main input of the experiment are sixteen topic networks, derived from the Rexplore database [17], representing the co-occurrences of topics in the 1995–2010 timeframe. From a practical perspective, each network can be represented as a fully weighted graph $G_{year} = (V_{year}, E_{year})$, in which V is the set of keywords while E is the set of links representing co-occurrences between keywords.

The node weight is given by the number of publications in which the keyword appears, while the link weight is equal to the number of publications in which two keywords co-occur together in a particular year. However, as pointed out in [19], the use of keywords as proxies for topics suffers from a number of problems. In fact some keywords tend to be noisy and do not represent topics (e.g., "case study") while multiple keywords can refer to the same topic (e.g., "ontology mapping" and "ontology matching"). To address this issue, we automatically transformed the graph of keywords into a graph of topics using an ontology of computer science produced by Klink-2 [18].

Klink-2 is an algorithm which analyses keywords and their relationships with research papers, authors, venues, and organizations and takes advantage of multiple knowledge sources available on the web in order to produce an ontology of research topics linked by three different semantic relationships. It was run on a sample of about 19 million papers, yielding an ontology including about 15000 topics in the field of Computer Science. We converted the keyword network to a topic network by filtering out all the keywords that do not represent topics and by aggregating the keywords representing the same concept. For example, we aggregated keywords such as "semantic web", "semantic web technology" and "semantic web technologies" in a single node and accordingly recomputed the weights of the network.

From the topic networks we selected two initial groups of topics. The first group, labelled debutant topics was composed by topics that made their debut in the period between 2000 and 2010. The second group, labelled control group or non-debutant group, included topics that made their debut long before the debutant ones (at least in the previous decade) and thus were already established when analysed.

As we will discuss in Sect. 4, we firstly conducted a preliminary evaluation while designing the approach, with the aim of choosing the best combination of technologies for this task. We then evaluated the method on a bigger sample of topics. In the preliminary phase, we focused only on the Semantic Web (debuting in 2001) and Cloud Computing (2006) as debutant topics, because they are well-known research areas and this facilitated the process of validation. For the non-debutant group we selected twenty topics. In the second evaluation, we randomly chose 50 topics for the debutant group and 50 topics for non-debutant group.

3.2 Selection Phase

The selection phase is the first step of this approach and, as already mentioned, it aims to select and extract portions of the collaboration networks related to topics in the two groups, in a few years prior to the year of analysis.

We hypothesised that after a new topic emerges it will continue to collaborate with the topics that contributed to its creation for a certain time. Hence, for each debuting topic we extracted the portion of topic network containing its n most co-occurring topics and analysed them in the five years preceding its year of debut. In brief, if a topic A makes its debut in 2003, the portion of network containing its most related topics will be analysed in the 1998–2002 time frame,

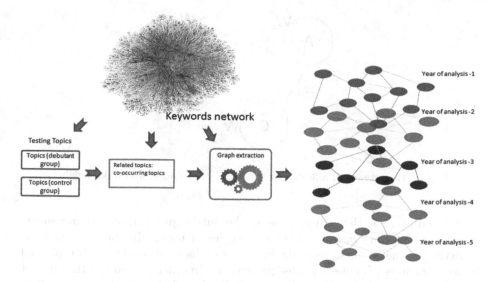

Fig. 1. Workflow representing all the steps for the selection phase

as showed in Fig. 1. We repeated the same procedure on the topics in the control group, assigning them a random year of analysis within the decade 2000–2010. We performed a number of experiments considering different values of n (20, 40, and 60).

At the end of the selection phase we associated to each topic in the two groups a graph G^{topic}:

$$G^{topic} = G^{topic}_{year-5} \cup G^{topic}_{year-4} \cup G^{topic}_{year-3} \cup G^{topic}_{year-2} \cup G^{topic}_{year-1} \qquad (1)$$

which corresponded to its collaboration network in the five years prior to its emergence. This graph contained five sub-graphs G^{topic}_{year-i} and each one corresponded to:

$$G^{topic}_{year-i} = (V^{topic}_{year-i}, E^{topic}_{year-i}) \qquad (2)$$

in which V^{topic}_{year-i} is the set of most co-occurring topics in a particular year and E^{topic}_{year-i} is the set of edges that link nodes in the set V^{topic}_{year-i}.

3.3   Analysis Phase

In this phase we evaluated the pace of collaboration between topics in the sub-graphs by analysing how the weights associated to nodes and links evolved in time. To this aim we transformed the graphs in sets of 3-cliques. A 3-clique, as shown in Fig. 2, is a complete sub-graph of order three in which all nodes are connected to one another and it is employed to model small groups of entities close to each other [13].

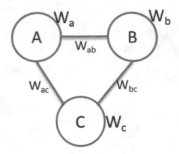

Fig. 2. An instance of a 3-clique containing both nodes, and links weights

The intuition is that we can assess the sub-graphs activity by measuring the increase of collaboration in these triangles of topics. In the first instance, we extracted the 3-cliques from the five sub-graphs associated to each topic and created timelines of cliques in subsequent years. In order to measure the amount of collaboration associated to a clique we devised the index showed in Formula 3, which measures the collaboration of nodes $\{A, B, C\}$ by taking in consideration both node weights $\{W_a, W_b, W_c\}$ and link weights $\{W_{ab}, W_{bc}, W_{ac}\}$. It does so by computing the conditional probability $P(y|x) = W_{xy}/W_x$ that a publication associated with a topic x will be also associated with a topic y in a certain year. The advantage of using the conditional probability over the number of co-occurrences is that the resulting value is already normalised according to the dimension of the topics.

$$
\begin{aligned}
\mu_1 &= harmmean\left(P(A|B), P(B|A)\right) \\
\mu_2 &= harmmean\left(P(B|C), P(C,B)\right) \\
\mu_3 &= harmmean\left(P(C|A), P(A|C)\right) \\
\mu_\Delta &= harmmean\left(\mu_1, \mu_2, \mu_3\right)
\end{aligned}
\tag{3}
$$

This approach computes the weight associated to each link between topic x and y by using the harmonic mean of the conditional probabilities $P(y|x)$ and $P(x|y)$ and then computes the final index μ_Δ as the harmonic mean of all the weights of the clique. We tested other kind of means (e.g., arithmetic mean) in the preliminary evaluation, but the harmonic mean appears to work better, as we will show in Sect. 4.1, since it rewards cliques in which all the links are associated with high values in both directions.

At this stage, each clique is now reduced to a timeline of measures, as showed in Formula 4. We then studied the evolution of these values to determine whether the collaboration pace of a clique was increasing or decreasing, as showed in Fig. 3.

$$
\mu_{\Delta time}^{clique-i} = \left[\mu_{(\Delta yr-5)}, \mu_{(\Delta yr-4)}, \mu_{(\Delta yr-3)}, \mu_{(\Delta yr-2)}, \mu_{(\Delta yr-1)}\right]
\tag{4}
$$

We first tried to determine the tendency of a clique by simply taking the difference between the first and the last values of the timeline. However, this method

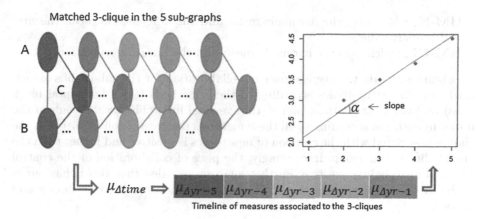

Fig. 3. Main steps of the analysis phase: from 3-cliques matching to slope processing

ignores the other values in the timeline and can thus ignore important information. For this reason, we applied the linear interpolation method on the five indexes using the least-squares approximation to determine the linear regression of the time series $f(x) = \alpha \cdot x + \beta$. The slope α is then used to assess the increase of collaboration in a clique. When α is positive the degree of collaboration between the topics in the clique is increasing over time, while if it is negative the topics are growing more distant. Subsequently, the collaboration pace of each sub-graph was assessed by computing the average and standard deviation of the slopes of the associated cliques.

4   Findings and Discussion

We will now report the results of the preliminary and full evaluation. The latter was performed on a dataset of 3 million publications including 100 topics initially selected for the analysis (50 debutant topics and 50 topics for the control group), and over 2000 of their co-occurring topics.

4.1   Preliminary Evaluation

In Sect. 3, we discussed two techniques to compute the weight of a clique (i.e., harmonic mean and arithmetic mean) and two methods to evaluate its trend (i.e., computing the difference between the first and the last values and linear interpolation). We tested these four techniques on the graphs composed by the 20 most co-occurring topics per each testing topics. In particular, we evaluated the following approaches:

- **AM-N**, which uses the arithmetic mean and the difference between the two extreme values;
- **AM-CF**, which uses the arithmetic mean and the linear interpolation;

- **HM-N**, which uses the harmonic mean and the difference between the first and the last values;
- **HM-CF**, which uses the harmonic mean and the linear interpolation.

Figure 4 reports the average pace of collaboration for the sub-graphs associated to each testing topics according to these methods (thick horizontal black lines) and the range of their values (thin vertical line). The results confirm the initial hypothesis: according to all these methods the pace of collaboration in the cliques associated with the creation of new topics is positive and higher than the one of the control group. Interestingly, the pace of collaboration of the control group is also slightly positive. Further analysis revealed that this behaviour is probably caused by the fact that in time the topic network becomes denser and noisier.

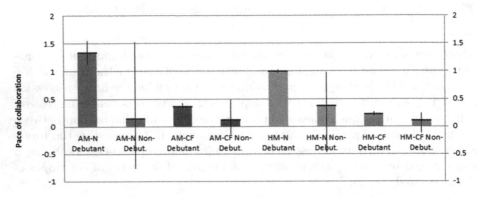

Fig. 4. Overall directions of the sub-graphs related to testing topics in both debutant and control group with all the four approaches.

The techniques based on the simple difference (AM-N and HM-N) exhibit the larger gap between the two groups in terms of average pace of collaboration. However, the ranges of values actually overlap, making it harder to assess if a certain sub-group is incubating a novel topic. The same applies to AM-CF. HM-CF performs better and even if the values slightly overlap when averaging the pace over different years they do not in single years. Indeed, analysing the two ranges separately in 2001 and 2006 (see Fig. 5), we can see that the overall collaboration paces of the debutant topics (DB) are always significantly higher than the control group (NDB).

We ran the Student's t-test on the HM-CF approach in order to verify that the two groups, showed in Fig. 6, actually belong to different populations and thus the initial hypothesis is supported by empirical evidence. The test yielded

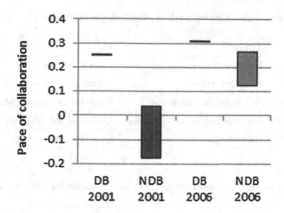

Fig. 5. Overall directions of the sub-graphs related to testing topics in both debutant and control group in HM-CF approach

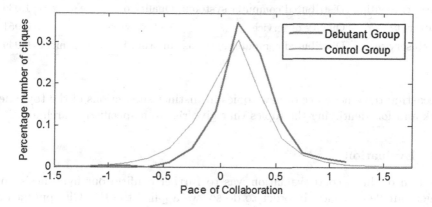

Fig. 6. Distributions of slope valued for both groups

a p-value equal to $7.0280 \cdot 10^{-12}$, which allows us to reject the null hypothesis that the differences between the two distributions are due to random variations.

The results of HM-CF show also interesting insights on the creation of some well-known research topics. Tables 1 and 2 list the cliques which exhibited a higher slope for semantic web and cloud computing. In particular, semantic web was anticipated in the 1996–2001 timeframe by a significant increase in the collaborations of the world wide web area with topics such as information retrieval, artificial intelligence, and knowledge based systems. This is actually consistent with the initial vision of the semantic web, defined in the 2001 by the seminal work of Tim Berners-Lee [2].

Similarly, cloud computing was anticipated by an increase in the collaboration between topics such as grid computing, web services, distributed computer systems and internet. This suggests that our approach can be used both for

Table 1. Ranking of the cliques with highest slope value for the "semantic web".

Topic 1	Topic 2	Topic 3	Score
World wide web	Information retrieval	Search engines	2.529
World wide web	User interfaces	Artificial intelligence	1.12
World wide web	Artificial intelligence	Knowledge representation	0.974
World wide web	Knowledge based systems	Artificial intelligence	0.850
World wide web	Information retrieval	Knowledge representation	0.803

Table 2. Ranking of the cliques with highest slope value for the "cloud computing".

Topic 1	Topic 2	Topic 3	Score
Grid computing	Distributed computer systems	Web services	1.208
Web services	Information management	Information technology	1.094
Grid computing	Distributed computer systems	Quality of service	1.036
Internet	Quality of service	Web services	0.951
Web services	Distributed computer systems	Information management	0.949

forecasting the emergence of new topics in distinct subsections of the topic network and for identifying the topics that give rise to a specific research area.

4.2   Evaluation

The aim of this second evaluation was to further confirm our hypothesis on a bigger sample of topics. In order to do so, we applied the HM-CF approach on 50 debutant topics and compared them to a control group of 50 non-debutant topics. In particular, we performed a number of tests varying the number of co-occurring topics selected per each testing topic.

The charts in Fig. 7 reports the results obtained by using 20, 40 and 60 co-occurring topics. Each bar shows the mean value of the average pace of collaboration for the debutant (DB) and non-debutant (NDB) topics. As before, the average pace computed in the portion of topic network related to debutant topics is higher than the one of the control group.

Figure 8 shows the average collaboration pace for each year when considering the 20 most co-occurring topics. The collaboration pace for the debutant topics is higher than the one for the control group with the exception of 2009, when they were almost equal. In addition, in the last five years the overall pace of the non-debutant topics fluctuates, while the overall directions for the debutant topics suffer a significant fall. This can be due to a variety of factors. First, as we mentioned before, the topic network became denser and noisier in recent years.

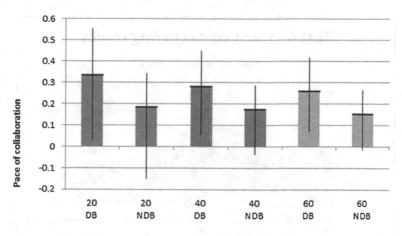

Fig. 7. Average collaboration pace of the sub-graphs associated to the debutant (DB) and control group (NDB) topics, when selecting the 20, 40 and 60 most co-occurring topics. The thin vertical lines represent the ranges of the values

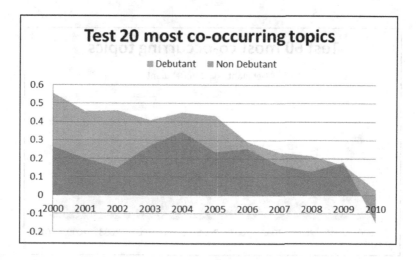

Fig. 8. Average collaboration pace per year of the sub-graphs related to testing topics in both debutant and control group considering their 20 most co-occurring topics. The year refers to the year of analysis of each topic

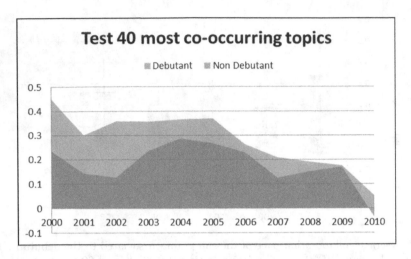

Fig. 9. Average collaboration pace per year of the sub-graphs related to testing topics in both debutant and control group considering their 40 most co-occurring topics

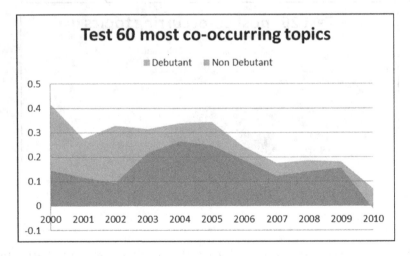

Fig. 10. Average collaboration pace per year of the sub-graphs related to testing topics in both debutant and control group considering their 60 most co-occurring topics

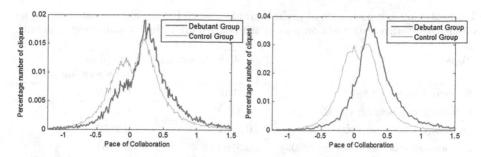

Fig. 11. Distributions of slopes in the year 2001 (left) and 2002 (right) when considering the 60 most co-occurring topics

Moreover, the most recent debutant topics often have a yet underdeveloped network of collaborations, which may results in a poor selection of the group of topics to be analysed in the previous years. Therefore, selecting only 20 most co-occurring topics may not allow us to highlight the correct dynamics preceding the topic creation.

Indeed, choosing a higher number of co-occurring topics significantly alleviates this issue. The effect is reduced when selecting 40 of them (Fig. 9) and with 60 the collaboration pace of debutant topics is always significantly higher than the one for the control group (Fig. 10). However, the fall in the last five years is still present and we thus intend to further investigate this phenomenon in future work.

We ran the Student's t-test on the groups in different years, in order to confirm that the two distributions belong to different populations. When taking in consideration the 20 most co-occurring topics, the Student t-test yields $p = 0.04$ in 2009 and $p < 1.36 \cdot 10^{-20}$ in other years, whereas, when taking 40 and 60 most co-occurring topics the p-values are all less than $1.28 \cdot 10^{-51}$. As an example, Fig. 11 shows the distributions in 2000 and 2001 for the 60 most co-occurring topics.

Table 3 shows a selection of debutant topics and their collaboration pace versus the collaboration pace of the control group in the same year. We can see a good number of well-known topics that emerged in the last decade and how their appearance was anticipated by the dynamics of the topic network.

In conclusion, the results confirms that the portions of the topic network in which a novel topic will appear exhibit a measurable fingerprint, in terms of increased collaboration pace, well before the topic is recognized and labelled by researchers. These dynamics can be exploited to foster the early detection of emerging research trends.

Table 3. Collaboration pace of the sub-graphs associated to selected debutant topics versus the average collaboration pace of the control group in the same year of debut.

Topic	Collaboration pace	Standard collaboration pace
Service discovery (2000)	0.4549	0.1459
Ontology engineering (2000)	0.4350	0.1459
Ontology alignment (2005)	0.3864	0.2473
Service-oriented architecture (2003)	0.3598	0.2164
Smart power grids (2005)	0.3580	0.2473
Sentiment analysis (2005)	0.3495	0.2473
Semantic web services (2003)	0.3493	0.2164
Linked data (2004)	0.3477	0.2638
Wimax (2004)	0.3470	0.2638
Semantic web technology (2001)	0.3434	0.1160
Vehicular ad hoc networks (2004)	0.3421	0.2638
Manet (2001)	0.3416	0.1160
P2P network (2002)	0.3396	0.0947
Location based services (2001)	0.3308	0.1160
Service oriented computing (2003)	0.3306	0.2164
Ambient intelligence (2002)	0.2892	0.0947
Social tagging (2006)	0.2631	0.1865
Wireless sensor network (2001)	0.2583	0.1160
Community detection (2006)	0.2433	0.1865
Cloud computing (2006)	0.2410	0.1865
User-generated content (2006)	0.2404	0.1865
Information retrieval technology (2008)	0.2315	0.1411
Web 2.0 (2006)	0.2241	0.1865
Ambient assisted living (2006)	0.2236	0.1865
Internet of things (2009)	0.2214	0.1556

5   Conclusions

In this paper, we theorize that it is possible to detect topics in their embryonic stage, i.e., when they have not yet been labelled or associated with a considerable number of publications, by analysing the dynamics between existent topics. We also introduced a method for assessing the increase in the pace of collaboration of topic cliques and used it to confirm our hypothesis by testing it on more than 2000 topics and 3 million research publications. In particular, we selected a number of debuting topics and analysed the behaviour of their most co-occurring topics in the five years before their debut. We found that the pace of collaboration is significantly higher than the one of the control group. We plan to further develop our approach in two main directions. First, we are currently working on a method for the automatic detection of embryonic topics that analyses the topic network and identifies sub-graphs where topics exhibit the discussed dynamics. A second direction of work focuses on improving the current approach by integrating a number of additional dynamics involving other research entities, such as authors and venues. The aim is to produce a robust approach that could be used by researchers and companies alike for gaining a better understanding of where research is heading.

Acknowledgements. We would like to thank Springer Nature (http://www.springer. com/gp/) for partially funding this research and Elsevier B.V. (https://www.elsevier. com/) for providing us with access to their large repositories of scholarly data.

References

1. Becher, T., Trowler, P.: Academic Tribes and Territories: Intellectual Enquiry and the Culture of Disciplines. McGraw-Hill Education, New York (2001)
2. Berners-Lee, T., Hendler, J., Lassila, O.: The semantic web. Sci. Am. **284**, 28–37 (2001)
3. Blei, D.M., Ng, A.Y., Jordan, M.I.: Latent dirichlet allocation. J. Mach. Learn. Res. **3**, 993–1022 (2003)
4. Bolelli, L., Ertekin, Ş., Giles, C.L.: Topic and trend detection in text collections using latent dirichlet allocation. In: Boughanem, M., Berrut, C., Mothe, J., Soule-Dupuy, C. (eds.) ECIR 2009. LNCS, vol. 5478, pp. 776–780. Springer, Heidelberg (2009). doi:10. 1007/978-3-642-00958-7_84
5. Decker, S.L., Aleman-Meza, B., Cameron, D., Arpinar, I.B.: Detection of bursty and emerging trends towards identification of researchers at the early stage of trends. University of Georgia (2007)
6. Duvvuru, A., Kamarthi, S., Sultornsanee, S.: Undercovering research trends: network analysis of keywords in scholarly articles. In: 2012 International Joint Conference on Computer Science and Software Engineering (JCSSE), pp. 265–270 (2012)
7. Duvvuru, A., Radhakrishnan, S., More, D., Kamarthi, S., Sultornsanee, S.: Analyzing structural & temporal characteristics of keyword system in academic research articles. Procedia Comput. Sci. **20**, 439–445 (2013)
8. Erten, C., Harding, P.J., Kobourov, S.G., Wampler, K., Yee, G.: Exploring the computing literature using temporal graph visualization. In: Electronic Imaging 2004, pp. 45–56 (2004)
9. Gruhl, D., Guha, R., Liben-Nowell, D., Tomkins, A.: Information diffusion through blogspace. In: Proceedings of the 13th International Conference on World Wide Web, pp. 491–501 (2004)
10. He, Q., Chen, B., Pei, J., Qiu, B., Mitra, P., Giles, L.: Detecting topic evolution in scientific literature: how can citations help? In: Proceedings of the 18th ACM Conference on Information and Knowledge Management, pp. 957–966 (2009)
11. Jo, Y., Lagoze, C., Giles, C.L.: Detecting research topics via the correlation between graphs and texts. In: Proceedings of the 13th ACM SIGKDD International Conference on Knowledge Discovery and Data Mining, pp. 370–379 (2007)
12. Kuhn, T.S.: The Structure of Scientific Revolutions. University of Chicago Press, Chicago (2012)

13. Luce, R.D., Perry, A.D.: A method of matrix analysis of group structure. Psychometrika **14**, 95–116 (1949)
14. Lv, P.H., Wang, G.-F., Wan, Y., Liu, J., Liu, Q., Ma, F.-C.: Bibliometric trend analysis on global graphene research. Scientometrics **88**, 399–419 (2011)
15. Mathioudakis, M., Koudas, N.: Twittermonitor: trend detection over the twitter stream. In: Proceedings of the 2010 ACM SIGMOD International Conference on Management of Data, pp. 1155–1158 (2010)
16. Morinaga, S., Yamanishi, K.: Tracking dynamics of topic trends using a finite mixture model. In: Proceedings of the Tenth ACM SIGKDD International Conference on Knowledge Discovery and Data Mining, pp. 811–816 (2004)
17. Osborne, F., Motta, E., Mulholland, P.: Exploring scholarly data with rexplore. In: Alani, H., et al. (eds.) ISWC 2013. LNCS, vol. 8218, pp. 460–477. Springer, Heidelberg (2013). doi:10.1007/978-3-642-41335-3_29
18. Osborne, F., Motta, E.: Klink-2: integrating multiple web sources to generate semantic topic networks. In: Arenas, M., et al. (eds.) ISWC 2015. LNCS, vol. 9366, pp. 408–424. Springer, Cham (2015). doi:10.1007/978-3-319-25007-6_24
19. Osborne, F., Motta, E.: Mining semantic relations between research areas. In: Cudré-Mauroux, P., et al. (eds.) ISWC 2012. LNCS, vol. 7649, pp. 410–426. Springer, Heidelberg (2012). doi:10.1007/978-3-642-35176-1_26
20. Osborne, F., Scavo, G., Motta, E.: A hybrid semantic approach to building dynamic maps of research communities. In: Janowicz, K., Schlobach, S., Lambrix, P., Hyvönen, E. (eds.) EKAW 2014. LNCS (LNAI), vol. 8876, pp. 356–372. Springer, Cham (2014). doi:10.1007/978-3-319-13704-9_28
21. Rosen-Zvi, M., Griffiths, T., Steyvers, M., Smyth, P.: The author-topic model for authors and documents. In: Proceedings of the 20th Conference on Uncertainty in Artificial Intelligence, pp. 487–494 (2004)
22. Salatino, A.: Early detection and forecasting of research trends (2015)
23. Sun, X., Kaur, J., Milojevi, S., Flammini, A., Menczer, F.: Social dynamics of science. Sci. Rep. **3**, 1069 (2013)
24. Tseng, Y.-H., Lin, Y.-I., Lee, Y.-Y., Hung, W.-C., Lee, C.-H.: A comparison of methods for detecting hot topics. Scientometrics **81**, 73–90 (2009)

Dynamic Visualization of Citation Networks Showing the Influence of Scholarly Fields over Time

Jason Portenoy[✉] and Jevin D. West

The Information School, University of Washington, Seattle, WA, USA
{jporteno,jevinw}@uw.edu

Abstract. Citation graphs between scholarly papers can be used to learn about the structure and development of scholarship. We present a generalizable approach to visualizing scholarly influence over time, using a dynamic node-link diagram representing the citation patterns between groups of papers. We combine this approach with hierarchical clustering techniques that exploit the network structure to partition the graph into clusters representing fields and subfields. We use these methods to explore the influence that fields have had on other fields over time.

Keywords: Big scholarly data · Citation networks · Dynamic network visualization · Scholarly evaluation · Science of science

1 Introduction

Examining the scholarly literature as a vast network of ideas connected by citations and footnotes can yield insights into the flow and evolution of ideas. Understanding the structure of this network can help us understand the progression of science and the development of human knowledge. We present novel methods using dynamic network visualizations and graph clustering techniques to show patterns around the influence that scholarly fields have on other fields over time.

The node-link diagram is a common paradigm for visually representing network data; however, these diagrams tend to be overwhelming and inscrutable to the average viewer, especially when representing rich dynamic data. Our approach is to use animation and a spiral placement of nodes around a fixed center to encode temporal features of the network, allowing for a more accessible display of rich information in a limited space.

2 Motivation and Background

We were motivated to explore visualization methods of scholarly influence through a collaboration with the Pew Scholars Program in the Biomedical Sciences, which provides early-career funding to prominent health researchers

A. González-Beltrán et al. (Eds.): SAVE-SD 2016, LNCS 9792, pp. 147–151, 2016.
DOI: 10.1007/978-3-319-53637-8_14

each year. The program wanted to reflect on its history using more than just standard metrics such as citation counts, h-indexes, and impact factors.

We approached our work with the Pew program as a case study in narrative visualization, developing our methods according to the Pew program's goal of reflecting on their history while seeing a broader opportunity to use visualization to convey scholarly impact. We drew on previous work in visualizing citation networks [3], narrative visualizations [5], and dynamic network visualizations [1] to provide ways of exploring dimensions of scholarly influence over time in an accessible and compelling format. We designed the visualization in collaboration with the Pew program leadership, and evaluated the design using demonstrations and interviews with Pew Scholars. A demonstration of the author-level visualization using sample data can be found at http://scholar.eigenfactor.org/demo. We concluded from our evaluation that our design was an effective method of showing scholarly influence.

In our current work, we explore shifting the focus from author to academic field. We partition the overall citation graph into fields and subfields (see Sect. 3.1) and focus on one specific subfield as the center node. Extending our efforts visualizing the influence that one scholar has had within and outside of her research area over time, we hope to explore the influence that an idea in science has had over the course of its lifetime.

3   Methods

3.1   Data

Our database of scholarly publications includes 1.7 million full-text articles from the JSTOR archive and more than 8 million citations between those articles. The methods for naming the fields are described in [8]. We used the citation network–with papers as nodes and citations as links–and clustered the citation graph using the hierarchical version of the map equation [4]. We then hand-labeled 1,765 fields and subfields from the hierarchical partition. We labeled the fields by looking at the top 50 papers in each field.

Papers were ranked using the Article-Level Eigenfactor Score (ALEF). The details of the ranking method can be found at [7]. It takes into account the time-directedness of article-level networks. The ALEF algorithm has performed well in static ranking challenges [6]. We used the article rankings for both the labeling and for identifying the influence of papers that cite the fields of focus (see Fig. 1).

Currently, the visualization samples 6 fields. We plan to expand to all 1,765 fields in the JSTOR dataset. Users will have the ability to move from field to field and from paper to paper. We plan also to cluster, label, and incorporate other paper archives, including SSRN, arXiv, etc.

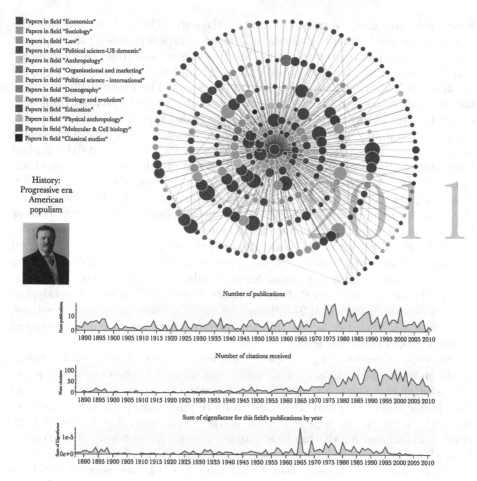

Fig. 1. An example of the visualization for the field of "Progressive era American Populism". The center node represents all publications within this field. Nodes that appear around the center represent publications that cited work in this field. Node size shows a citation-based indicator (Eigenfactor [7]) of the paper's influence. Node color shows the high-level field of each paper (e.g. "Economics," "Sociology"). Integrated timeline charts below show number of publications in the field of focus, number of citations received by that field, and the sum of the Eigenfactor for all of that field's papers.

3.2   Visualization Design

Figure 1 shows a screenshot of the final state of the visualization; the full, animated, interactive visualization for a sample of fields can be viewed at http://scholar.eigenfactor.org/fields. We use the open-source JavaScript visualization library D3 [2], transforming the citation data into a directed egocentric network in which the center (ego) node represents all of the publications in a particular subfield and surrounding (alter) nodes represent individual papers from other fields that have cited work represented in the ego. (The code to generate the visualization will be made available in an open-source repository.) In the figure above, for example, the central subfield is "Progressive era American populism" which is part of the high-level field "History," and the alter nodes are papers in other subfields that have cited this field's papers. The graph diagram shows the ego node as a central circle, and the alter nodes as circles that surround the center in a spiral formation. The surrounding nodes appear one by one according to year of publication and send out links representing citations to the center and to other nodes that appear in the network. Papers that cite the central field multiple times send out multiple overlaid links to the center, so these links appear darker. A year counter shows the publication year of the papers currently appearing. As the nodes appear, the viewer can explore further by mousing over the nodes for more information. The viewer can also explore papers in the center field by mousing over the timeline charts below.

While the total set of the nodes represents any paper that has ever cited a paper authored by the central field, in order to reduce the complexity of the graph we choose a sample of these nodes to visualize, aiming to include influential papers (based on Eigenfactor), from a variety of high-level fields. Currently, the number of nodes is capped at 275, though we are experimenting with alternatives. More complete statistics for the full egocentric network are shown in integrated timeline charts below the graph.

We explore representing the idea of influence in several different ways, and so the network diagram features several different visual encodings of influence. The *size of the nodes* is scaled by the Eigenfactor metric of each paper, so that larger nodes are easily identified as more influential papers. The *color of the nodes* encodes the high-level field of the paper according to the paper's cluster assignment for the top-level partition (e.g. "Economics," "Sociology"). This allows a view of to what extent this subdiscipline has had impact that has spread to other disciplines. A field with a more monochrome network will have had most of its influence within one particular field, while a network with more color means more citations from papers in other fields.

We arrived at our design through an iterative design process with the overall goal of creating an accessible, narrative visualization of different dimensions of scholarly influence over time. The use of animation was an important design choice throughout the process, as animation naturally draws attention and can encourage perceptions of narrative. We chose the spiral for the spatial encoding as it allowed us to encode time of publication as radial distance from the center, reinforcing its temporal encoding with the animation. The spiral layout also allows us to include more nodes in a limited space without too much overlap and confusion, addressing the overwhelming "hairball" effect that often comes along with node-link network diagrams.

4   Conclusions and Future Directions

Visualizing the citation graph in this way with the focus on a particular subfield gives a gestalt view of the spread of influence the field has had over time.

The work is in early stages, so we are still exploring what insights can come from examining different fields. Looking at the example in Fig. 1, we can see that "Progressive era American populism" is a well established field with a history of publication going back to the 1800s. As may be expected, it has had a wide range of influence across many academic disciplines, as can be seen by the extent of the colors in the network diagram.

We plan to use this work as a starting point to explore different ways of visualizing dynamic field-level influence using the citation network. We currently make use of the top and bottom levels of the cluster hierarchy, but we can explore ways to make use of the intermediate levels. We can also make use of other data, such as marking certain nodes as review articles, as these may have a different interpretation than primary research papers when thinking about influence. Another direction would be to switch the citation direction and, rather than looking at the influence that a field has had, visualize which papers and fields have influenced the field of interest.

Acknowledgements. We would like to thank the Pew Charitable Trusts and the Metaknowledge Network for funding and other support. We also thank JSTOR and Microsoft Academic Search for allowing us the use of their data.

References

1. Beck, F., Burch, M., Diehl, S., Weiskopf, D.: The state of the art in visualizing dynamic graphs. In: EuroVis STAR (2014)
2. Bostock, M., Ogievetsky, V., Heer, J.: D3: data-driven documents. IEEE Trans. Vis. Comput. Graph. **17**, 2301–2309 (2011). (Proc. InfoVis)
3. Cobo, M.J., López-Herrera, A.G., Herrera-Viedma, E., Herrera, F.: Science mapping software tools: review, analysis, and cooperative study among tools. J. Am. Soc. Inf. Sci. Technol. **62**(7), 1382–1402 (2011). doi:10.1002/asi.21525
4. Rosvall, M., Bergstrom, C.T.: Multilevel compression of random walks on networks reveals hierarchical organization in large integrated systems. PLoS ONE **6**(4), e18209 (2011). doi:10.1371/journal.pone.0018209
5. Segel, E., Heer, J.: Narrative visualization: telling stories with data. IEEE Trans. Vis. Comput. Graph. **16**(6), 1139–1148 (2010). doi:10.1109/TVCG.2010.179
6. Wesley-Smith, I., Bergstrom, C.T., West, J.D.: Static ranking of scholarly papers using article-level eigenfactor (ALEF). ACM (2016)
7. West, J.D., Bergstrom, T.C., Bergstrom, C.T.: The eigenfactor MetricsTM: a network approach to assessing scholarly journals. Coll. Res. Libr. **71**(3), 236–244 (2010). doi:10.5860/0710236
8. West, J.D., Jacquet, J., King, M.M., Correll, S.J., Bergstrom, C.T.: The role of gender in scholarly authorship. PloS One **8**(7), e66212 (2013)

Author Index

Printed in the United States
By Bookmasters